AF713593

Workplace Safety and Health Practices

Safety and health concerns in workplaces across the globe continue to be of concern for leaders as they seek to discover novel ways to provide a workplace that preserves the health and safety of workers. To achieve world-class health and safety in any organization, the health and safety of workers must be integrated with technological innovations that monitor, track, and provide feedback on worker performance and safety. These innovations are expected to be comfortable, smart, and even fashionable to achieve increased compliance.

Workplace Safety and Health Practices investigates the current and future trends that are leading the way and improving workplace health and safety all over the world. Organized into four parts to provide ease of reading this book focuses on program administration and management engagement and considerations of human aspects for safe and inclusive workplaces. It looks into the identification, mitigation, and control of hazards found at work and the program challenges that are faced along with the development and implementation of improvement measures. The authors provide considerations for novel ways of taking the protection of workers to a higher level considering the evolving definition of the workplace and locations where work takes place. With case studies backing up the theories, the reader will develop a thorough and complex overview of what is working in global workplaces to deliver top-quality health and safety.

This book will be a fascinating read for any health and safety professional. Its appeal will extend to industrial hygiene professionals, system engineers, construction engineers, and ergonomics and human factors practitioners.

Sustainable Improvements in Environment Safety and Health

Series Editor: Frances Alston, Lawrence Livermore National Laboratory, California, USA

Lean Implementation

Applications and Hidden Costs
Frances Alston

Safety Culture and High-Risk Environments

A Leadership Perspective
Cindy L. Caldwell

Industrial Hygiene

Improving Worker Health through an Operational Risk Approach
Frances Alston, Emily J. Millikin, and Willie Piispanen

The Legal Aspects of Industrial Hygiene and Safety
Kurt W. Dreger

Strategic Environmental Performance

Obtaining and Sustaining Compliance
Frances Alston and Brian Perkins

Occupational Exposures

Chemical Carcinogens and Mutagens
Frances Alston and Onwuka Okorie

Industrial Hygiene

Improving Worker Health through an Operational Risk Approach, Second Edition
Frances Alston and Emily J. Millikin

Workplace Safety and Health Practices

Current and Evolving Trends
Frances Alston and Onwuka Okorie

For more information about this series, please visit: https://www.crcpress.com/Sustainable-Improvements-in-Environment-Safety-and-Health/book-series/CRCSUSIMPENVSAF

Workplace Safety and Health Practices

Current and Evolving Trends

Frances Alston and Onwuka Okorie

CRC Press
Taylor & Francis Group
Boca Raton London New York

CRC Press is an imprint of the
Taylor & Francis Group, an **informa** business

First edition published 2026
by CRC Press
2385 NW Executive Center Drive, Suite 320, Boca Raton FL 33431

and by CRC Press
4 Park Square, Milton Park, Abingdon, Oxon, OX14 4RN

CRC Press is an imprint of Taylor & Francis Group, LLC

© 2026 Frances Alston and Onwuka Okorie

Reasonable efforts have been made to publish reliable data and information, but the author and publisher cannot assume responsibility for the validity of all materials or the consequences of their use. The authors and publishers have attempted to trace the copyright holders of all material reproduced in this publication and apologize to copyright holders if permission to publish in this form has not been obtained. If any copyright material has not been acknowledged please write and let us know so we may rectify in any future reprint.

Except as permitted under U.S. Copyright Law, no part of this book may be reprinted, reproduced, transmitted, or utilized in any form by any electronic, mechanical, or other means, now known or hereafter invented, including photocopying, microfilming, and recording, or in any information storage or retrieval system, without written permission from the publishers.

For permission to photocopy or use material electronically from this work, access www.copyright.com or contact the Copyright Clearance Center, Inc. (CCC), 222 Rosewood Drive, Danvers, MA 01923, 978-750-8400. For works that are not available on CCC please contact mpkbookspermissions@tandf.co.uk

Trademark notice: Product or corporate names may be trademarks or registered trademarks and are used only for identification and explanation without intent to infringe.

ISBN: 978-1-032-61871-5 (hbk)
ISBN: 978-1-032-62437-2 (pbk)
ISBN: 978-1-032-62438-9 (ebk)

DOI: 10.1201/9781032624389

Typeset in Times
by Deanta Global Publishing Services, Chennai, India

Contents

Preface xiii
About the Authors xv

PART I

Chapter 1 Leadership Safety and Health Responsibilities 3

1.1 Leadership and Management Role in Safety Program Implementation and Improvement 3
1.1.1 Leadership and Management Role in Organization Development and Implementation 6
1.2 Competent Safety Leadership 8
1.2.1 What Does It Look Like? 9
1.2.2 Training Safety Leaders for the Present and the Future 12
References 15

Chapter 2 Leadership Selection and Development 16

2.1 Selecting and Developing Leadership from Within 16
2.1.1 How to Grow Leaders from Within 17
2.1.2 Developing Leadership Skills 18
2.1.3 First a Follower and Then a Leader 19
2.1.4 Selecting Leaders 21
2.2 Selecting and Hiring Leaders External to the Company 22
2.3 Leadership Ethics: Why Is It Important? 23
References 26

Chapter 3 Determining Regulatory Applicability 28

3.1 Evaluating the Business Risk Profile 28
3.1.1 Employment Laws Facilitation of Worker Safety and Health 29
3.2 Workplace Safety and Health Laws 31
3.2.1 Health and Safety Laws and Regulations 31
3.2.2 Environment Laws and Regulations 32
3.3 Ensuring and Maintaining Compliance 33
3.3.1 Assessment Planning 34
3.3.2 Perform the Compliance Assessment 35
3.3.3 Analyze Findings 35

3.3.4 Conduct Gap Analysis ... 35
3.3.5 Prepare Report and Communicate Results ... 36
3.3.6 Develop and Implement Improvement Plan ... 36
3.4 Benefits of Regulatory Compliance ... 36
3.4.1 Compliance Culture ... 37
3.4.2 Failure of Determining Regulatory Compliance ... 38
3.5 Whistleblowers ... 39
References ... 39

PART II

Chapter 4 Human Performance and Behavior Modification ... 43

4.1 The Impact of Behavior on Workplace Safety ... 43
4.1.1 People Make Errors ... 44
4.1.2 Human Characteristics ... 45
4.2 Work Culture Behavioral Impacts ... 47
4.2.1 Leadership and Managers' Behavioral Impacts ... 47
4.3 Behavior Modification Techniques ... 48
4.4 Human Performance Improvement Program ... 52
4.4.1 Understanding Error Precursors ... 52
References ... 53

Chapter 5 Culture Evaluations and Modifications ... 55

5.1 Organizational Culture ... 55
5.2 Building and Maintaining a High-Performance Organization ... 55
5.3 Building a High-Performance Culture ... 56
5.3.1 Top Talent Retention ... 57
5.3.2 Improved Performance ... 60
5.3.3 Customer Satisfaction ... 63
5.4 Maintaining a High-Performance Culture ... 64
5.4.1 Evaluating Organizational Culture ... 64
5.4.2 Employee Engagement ... 67
5.4.3 Leadership Role and Engagement ... 68
5.5 Developing and Implementing a Culture Improvement Plan ... 69
References ... 69

Chapter 6 The Art of Effective Communication ... 70

6.1 Communication of Hazard Analysis and Resultant Changes ... 70
6.2 Are You a Good Communicator? ... 71

6.3 Data Reporting and Communication 72
6.3.1 Communication with Employees 74
6.3.2 Communication with Supervisors 75
6.3.3 Communication with Senior Leaders 75
6.3.4 Communication with the Occupational Medicine Provider 75
6.4 Types of Scenarios Communicated by Safety and Health Professionals 78
6.5 Potential Overexposure Based on Monitoring Data 78
6.6 I Smell Something in My Work Area 81
References 83

PART III

Chapter 7 Hazard Analysis and Documentation 87

7.1 Why Conduct Hazard Analysis 87
7.2 Job Hazard Analysis (JHA) 88
The Basic Stages of a Job Hazard Analysis: 89
7.3 JHA Informed Exposure Assessment 91
7.4 Process Review and Analysis 91
7.5 Job Hazard Analysis Process Redesign 92
7.5.1 Post-Job Review of JHAs 92
7.6 Job Hazard Analysis Program 93
7.6.1 JHA Data Collection 94
7.6.2 Data Analysis 94
7.6.3 Data Interpretation 94
7.7 Prioritizing Work Tasks 95
7.8 How to Establish JHA Preventive Actions 96
7.8.1 Eliminate the Hazard 96
7.8.2 Contain the Hazard 96
7.8.3 Revise Work Procedures 96
7.8.4 Limit the Exposure 96
7.8.5 PPE Selection 96
7.9 Job Hazard Analysis Software 97
References 99

Chapter 8 Non-traditional Employee Safety and Health 101

8.1 Non-traditional Employees Safety and Health 101
8.1.1 Non-traditional Employment Advantages and Disadvantages 101
8.1.2 Employment Market Evolution 102
8.2 Gig Employees Safety and Health 104

8.3 Classification of Gig Workers ... 104
8.3.1 Safety Training ... 105
8.3.2 Personal Protective Equipment (PPE) ... 106
8.3.3 Reporting Safety Issues ... 106
8.3.4 Safety Enforcement ... 106
8.3.5 Access to Employer Resources ... 106
8.3.6 Sedentary Activity ... 106
8.3.7 Prone to Injuries ... 107
8.4 Telework Employees Safety and Health ... 107
8.5 Independent Contract Employees Safety and Health ... 111
8.5.1 Working for a Company that Follows OSHA ... 112
8.5.2 General Contractor Requirements ... 112
References ... 113

Chapter 9 Emergency Service Workers Health and Safety ... 114

9.1 Emergency Responders ... 114
9.1.1 History of Emergency Services ... 115
9.1.2 Firefighters and Emergency ... 115
9.1.3 United States Fire Administration ... 117
9.1.4 National Fire Protection Association ... 117
9.2 Emergency Service Worker ... 119
9.2.1 Basic Emergency Services ... 119
9.2.2 Specialized Emergency Services ... 119
9.2.3 Emergency Services Regulations ... 121
9.3 Emergency Medical Services Safety and Health ... 121
9.3.1 Engineering Controls for Emergency Medical Technicians (EMT) and Paramedics ... 122
9.3.2 Administrative Controls for Emergency Medical Technicians (EMT) and Paramedics ... 122
9.3.3 Personal Protective Equipment (PPE) ... 123
9.4 Emergency Responders Burnout ... 123
9.4.1 Emergency Responder Self-Care Techniques ... 124
9.5 Specialized Emergency Services Workers ... 124
9.5.1 Animal Control Services ... 125
References ... 130

Chapter 10 Measuring Performance ... 131

10.1 How Important Are Metrics in Telling the Story? ... 131
10.2 Metric Components and Types ... 131
10.2.1 Leading Indicators ... 132
10.2.2 Lagging Indicators ... 133
10.3 Why Should an Organization Track Performance Metrics? ... 133

10.4 Measuring Performance 134
10.5 How to Measure Safety and Health Performance 135
10.5.1 Examples and Interpretation of Safety and Health Metrics 137
10.6 Developing Performance Metrics That Demonstrates Actual Performance 141
10.7 Metrics a Facilitator of Continuous Improvement 143
10.7.1 Using Lessons Learned to Enhance Continuous Improvement 144
References 146

Chapter 11 Assessing and Auditing Safety and Health Program Effectiveness 147

11.1 Why Assess the Safety and Health Program? 147
11.2 Evaluating the Safety and Health Program 149
11.3 Developing and Selecting an Assessment Platform and Strategy 150
11.4 Implementing an Assessment Plan 152
11.5 Internal Program Assessment Strategies 153
11.5.1 Selection of the Assessment Team 154
11.6 External Assessments Team 154
11.7 Guidelines for Documentation of Assessments 155
11.8 Inspection Checklists 156
11.9 The Assessment In Brief (Kick Off) 156
11.10 The Assessment Out Brief (Closing) 157
11.11 Improvement Implementation Plan (IIP) 157

PART IV

Chapter 12 The Use of Technology to Improve Worker Safety and Health 161

12.1 Wearable Technology 161
12.1.1 Wearable Technology Products 162
12.1.2 Industrial Wearables Are Usually Divided into Four Groups Based on Function 162
12.1.3 Positive Safety Culture Supports Wearable Technology 163
12.1.4 Wearable Technology in Safety and Health 163
12.2 Adjustable Workstations 165
12.2.1 Custom Workstations 166
12.2.2 Types of Workstations 167
12.2.3 Factors to Consider for an Office Workstation 168
12.2.4 Workstation Safety 169

12.3 The Use of Drones ... 169
12.3.1 Drones Use in Improving Worker Safety and Health ... 170
12.3.2 Federal Aviation Administration ... 171
12.4 Visualization Technology ... 172
12.4.1 Visual Technology Used to Improve Workplace Safety ... 173
References ... 174

Chapter 13 Current Challenges for Safety and Health Workers ... 175

13.1 Protecting Vulnerable Workers ... 175
13.1.1 Workers with Disabilities ... 177
13.1.2 Pregnant Women ... 177
13.1.3 Workers Who Have Limited Knowledge and Experience ... 178
13.1.4 Workers That Have Different Cultural Experiences ... 178
13.1.5 Workers Having a Language Barrier ... 179
13.1.6 Workers New to the Workforce ... 179
13.1.7 Overworked Workers ... 180
13.1.8 Migrant Workers ... 180
13.2 Recognizing and Preventing Workplace Violence ... 182
13.3 Maintaining Workplace Equity ... 183
13.4 Workplace Laws and Regulations ... 183
13.4.1 Occupational Safety and Health ... 184
13.4.2 Department of Labor ... 184
References ... 186

Chapter 14 Barriers to Safety and Health ... 187

14.1 Safety and Health Program ... 187
14.1.1 Lack of Employee Engagement ... 187
14.1.2 Time Constraints ... 188
14.1.3 Budget Constraints ... 188
14.1.4 Inadequate Resources ... 189
14.1.5 Expertise ... 189
14.1.6 Different Regulations ... 189
14.1.7 Human Factors ... 189
14.1.8 Lack of Training ... 189
14.1.9 Poor Safety Culture and Practices ... 190
14.1.10 Worker Behavior ... 190
14.2 Exposure Surveillance ... 190
14.2.1 Exposure Hazard Monitoring ... 191
14.2.2 Health Monitoring ... 193
14.3 Improving Exposure and Risk Assessment ... 193

14.4 Certifying Exposure Control ... 194
14.4.1 Exposure Control Plan ... 195
14.4.2 Exposure Control Technologies ... 196
14.5 Proper Assessment and Controls during Pandemics ... 198
References ... 200

Chapter 15 Resource Packet ... 201

Resource Packet ... 201
Section 1 ... 201
Organization Culture Questionnaire ... 201
Culture Focus Group Questions ... 201
Employee Engagement Questionnaires ... 201
Organization Trust Questionnaire ... 202
Organization Diagnostic Questionnaire Matrix – Trust ... 202
Communication Survey Questions ... 202
Employee Engagement Questionnaire ... 205
Organization Trust Questionnaire ... 207
Organization Diagnostic Questionnaire Matrix – Trust ... 209
Organization Diagnostic Questionnaire Matrix – Trust - continued ... 209
Questions that can be included in a communication survey or round table ... 210
Section 2 ... 211
Exposure Assessment Data Collection Worksheet ... 212
AIR SAMPLING DATA SHEET ... 213
WORK AND ENVIRONMENTAL CONDITIONS ... 213
EQUIPMENT CALIBRATION DATA ... 214
SAMPLING RESULTS (ANALYSIS) ... 214
ANALYTICAL LABORATORY ... 214
PERSONAL PROTECTIVE EQUIPMENT (PPE) HAZARD ASSESSMENT CERTIFICATION AND SELECTION ... 214
Noise Hazard: (noise hazards greater than 85 dBA) ... 215
Eye Hazard: ... 215
Hand Hazard: ... 215
Foot Hazard: ... 215
Fall Hazard: ... 216
Respiratory Hazard: ... 216
Certified by: ... 216
SAMPLE JOB HAZARD ANALYSIS (JHA) ... 217
JHA Participants ... 217
Section 3 ... 217
Facility Basic Safety Conditions ... 218
Slips Trips and fall hazards ... 219
General Office Safety ... 220

Hazard Communication 220
Safety & Health Training 222
Respiratory Protection 222
Confined Space 223
Medical Surveillance 224
Ergonomics 225
Occupational Noise 226
ASSESSMENT PLAN EXAMPLE 227
Scope: 227
The Assessment Team 227
The Assessment Method 227
Data Collection 228
Results Reporting 228
Example of Criterion and Lines of Inquiries (LOIs) 228
Improvement Plan – Example 230
Section 4 231
Other Resources 231

Index 233

Preface

Safety and health concerns in workplaces across the globe continue to be of concern for leaders as they seek to discover novel ways to provide a workplace that preserves the health and safety of workers. As the world continues to evolve in diversity and types of products and technologies are developed, the field of safety and health too must continue to evolve. This evolution of worker protection results from the potential hazards associated with the development of new products and technologies as well as the repositioning of current products and technologies. During this evolution, practitioners and leadership must turn to and embrace creative technological changes that lend itself to enhancing the field and facilitating a safe work environment.

The workforce of today is what some may call non-traditional as workers are not always going to the office or in an industrial setting to conduct business. Non-traditional work is now increasingly found in certain sectors, such as information technologies, human resources, finance, and engineering. There are a lot of factors leading to the emergence of non-traditional workers. Some of these factors include the needs of the worker for flexible schedule, work-family balance, and their ability to maintain control of work-related decisions. The employer may not want to support the increasing cost associated with employee benefits and retain the flexibility to easily reduce costs, when necessary, through reducing the brick-and-mortar office environment. There is a perception that the younger generation is more willing than the older generation to work in non-traditional work arrangements. This increases the possibility that non-traditional work will continue to grow as the workforce changes over time.

Many white-collar and some blue-collar workers are working from their home offices and in some cases their automobiles. This practice has been magnified as a result of the COVID-19 outbreak where people distancing became the norm to avoid or limit exposure and therefore, transmitting the disease. To further complicate worker protection, many employees are a part of the gig workforce that can be engaged in a variety and types of jobs on a freelancing or full-time basis for a variety of companies. The question is who is responsible for the safety and health of these workers while they are working on the various projects? How can these workers protect themselves from the potential work hazards that they may be exposed to? The overarching question is one of accountability.

A safe workplace can become a competitive advantage for companies as workers choose their employers and work environment. Workers are seeking and are interested in working in environments that they perceive as not harmful to their health. Developing and assessing the effectiveness of these programs continues to be challenging for employers because some employers view these programs as costly and may not provide the return on investment that they seek.

Many of the safety and health challenges faced by workers and employers continue to expand into new territories. Therefore, non-traditional measures must be considered, and safety and health professionals must expand their knowledge and viewpoints of methods that can be effective in protecting workers. This change in viewpoints must include leveraging the various technologies that are available such as the use of drones to collect data and monitor areas that may pose a hazard to workers and supporting the use of wearable technologies to monitor workers while performing work. Such innovations in personal protective equipment now include equipment that monitors and collects data on heart rate, blood oxygen levels, blood pressure, and sweat levels. The use of embracement of innovative technologies is expected to improve job performance among workers and reduce injuries in the work environment.

This book is organized into four parts to provide ease of reading and to address systematic issues that are of strategic importance for safety and health leaders and professionals to consider in their efforts to provide safe workplaces for employees. Part I focuses on program administration and management engagement. This section is designed to assist in the development and implementation of an effective safety and health program. Part II focuses on the human aspects that must be considered and channeled in developing and implementing safety and health programs that provide the culture that underpins a safe and inclusive workplace. Part III primarily focuses on the identification, mitigation, and control of hazards found in the workplace. Part IV provides details of some of the program challenges that are faced along with the development and implementation of improvement measures, in addition, a resource packet containing tools that can be used in evaluating program effectiveness.

The authors attempt to provide considerations for novel ways of taking the protection of workers to a higher level considering the evolving definition of the workplace and locations where work actually takes place. For some workers, the workplace is in a traditional setting such as in an office, factory, or industrial setting. However, for other workers, the workplace is a location within the home or even their automobiles. Protecting workers that are working in diverse environments takes on new challenges that have not been at the forefront as today. In order to achieve world-class health and safety in any organization, the health and safety of workers must be integrated with technological innovations that monitor, track, and provide feedback on worker performance and safety. These innovations are expected to be comfortable, smart, and even fashionable in order to achieve increased compliance.

About the Authors

Dr Frances Alston has more than 30 years of leadership experience in high hazards and diverse work cultures. She is an instructor at a major university, a senior manager with an international company, and the Past President (2018) of the American Society for Engineering Management. Dr Alston is also a Fellow of the American Society for Engineering Management (ASEM) and has built a solid career leading the development and management of Environment, Safety, Health, and Quality (ESH&Q) programs in challenging cultural environments. She has a PhD in Industrial and System Engineering and an MSE degree in Engineering Management, both from the University of Alabama. She earned a Master's degree in Hazardous and Waste Materials Management/Environmental Engineering from Southern Methodist University and a Bachelor of Science Degree in Industrial Hygiene and Safety/Chemistry from Saint Augustine's University. She holds certifications as a Certified Hazardous Materials Manager (CHMM) and a Certified Professional Engineering Manager (CPEM).

Dr Onwuka Okorie currently works with the State of California. He has an extensive background in the field of industrial hygiene and occupational safety. He has worked with all levels of leadership in organizations toward achieving health and safety compliance and effective outcomes. He has a PhD in Occupational and Environmental Health and an MS degree in Occupational Health both from the University of Utah. He also has a Medical degree and a Postgraduate Medical Diploma. Dr Okorie is an AIHA Registered Specialist in Exposure Decision Analysis. He holds a certification in Project Management (PMP) and is a Certified Industrial Hygienist (CIH), a Certified Safety Professional (CSP), and a Fellow of the Royal Society for Public Health (RSPH).

Part I

Chapters 1–3 focus on program administration and management engagement. This part consists of three chapters that are designed to assist in the development and implementation of an effective safety and health program.

DOI: 10.1201/9781032624389-1

1 Leadership Safety and Health Responsibilities

1.1 LEADERSHIP AND MANAGEMENT ROLE IN SAFETY PROGRAM IMPLEMENTATION AND IMPROVEMENT

The role of managers and leaders in organizational development and maintenance is critical in ensuring the viability and sustainability of an organization and a company. Without leaders and managers organizations will not be able to accomplish their goals and successfully compete in their respective business markets. As such, these resources are and should be selected carefully due to the challenges that they face and skills required in implementing their roles as well as the importance of their role in the success of business ventures. To effectively engage the two roles, there must be a clear understanding of the two roles as they are and can be different yet important for the success of an organization. It has been shown based on experience that good managers do not necessarily make good leaders and vice versa. This concept is difficult for some to grasp because the two concepts are often used interchangeably. Although managers and leaders have distinct roles in organizations, there are some shared characteristics and responsibilities that one would expect to see to achieve success in goal implementation and fostering a safe work environment. These characteristics include but are not limited to commitment to the goals and business strategy, open and honest communication, integrity in actions and transactions, decisiveness in decision making, respect for others, confidence in abilities and capabilities to perform their role, and a creative and optimistic approach to accomplishing work activities.

Table 1.1 attempts to clarify and demonstrate the distinctions between the two roles.

An important aspect of the role of a leader is to keep the business afloat and prosperous through the development and enactment of a strategy (leadership responsibility generally starts with a vision and a strategy) and develop and implement policies and procedures to not only support the business process but to also support the safety and well-being of the workers they employ to assist with conducting their business. Managers are key in managing the resources and ensuring the goals and objectives of the business are met. This includes ensuring that the systems and processes that are in place are appropriate to achieve business success and sustainability. The focused responsibilities of managers and leaders are shown in Figures 1.1 and 1.2. The two figures show the ongoing iterative roles that leaders and managers engage in daily to keep the business afloat and ensure sustainability of a company or process. The roles are not always performed in any particular order once the business is up and running.

DOI: 10.1201/9781032624389-2

TABLE 1.1
Management and Leadership Contrast

Leadership Role	Management Role
Deals with the future through, for example, strategic planning	Deals with the present through, for example, guiding work activities and collaboration to support the implementation of business strategies
Inspires, motivates, and empowers others	Directs, executes, and monitors activities
Seeks after innovation	Seeks what works in support of the current business practices. May make changes in support of the current business
Sets the vision	Sets goals, plans, organizes, and delegates to make the vision a reality
People focused	Goals and objectives, processes, and systems focused
Sets the tone for culture	Provides structure for teams to support the culture

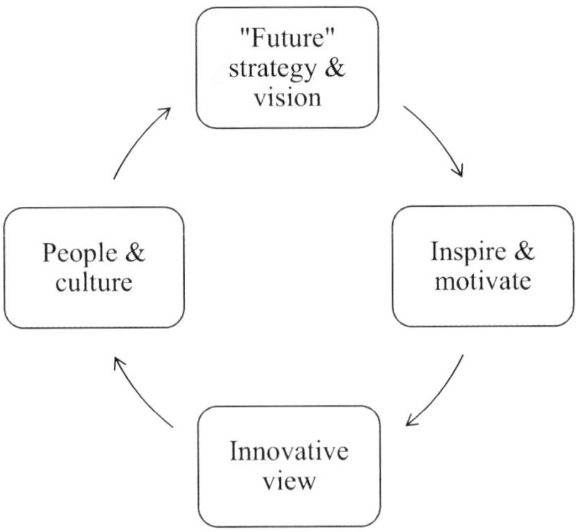

FIGURE 1.1 Leadership focused roles

Figure 1.1 represents the role that leadership predominately plays in an organization's success. One of the primary focuses of a leader is to visualize the future of the company and devise a strategy that will take the company into the future and sustain it in that futuristic environment. Another vital role of a leader is having the capability to inspire, influence, and motivate others to follow and utilize their knowledge and creative abilities to assist in vision realization. Leaders approach their role through an innovative lens that provides the basis for companies that need to distinguish themselves from their competitors. Leaders take responsibility for and pay close attention to the people and culture-building aspects of the business. They realize that culture and people are critical attributes that must be nurtured to build and sustain an organization and to facilitate a safe and productive work environment.

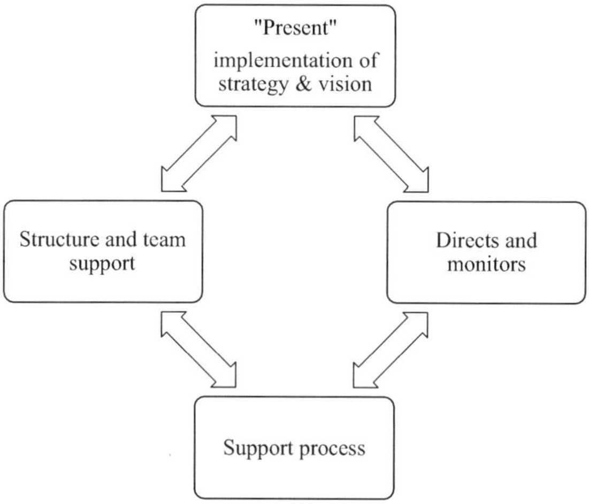

FIGURE 1.2 Management-focused roles

Figure 1.2 represents the role that management predominately plays in the success of an organization. A significant responsibility of management is to provide the building blocks and tools needed to support the implementation of the strategy and vision for the organization. To accomplish work managers are highly engaged in directing and monitoring work activities, supporting current processes that are in place modifying as needed to render the processes safe, effective, and efficient, and assembling and working with teams pulling together a collage of knowledge and talent to accomplish tasks and develop new processes, products, or systems.

The leadership and management-focused roles are documented and implemented through assuming responsibility for organization development and maintenance, developing, and implementing an organization structure that guides the way work is performed and accountability, development and implementation of policies and procedures, and the people aspects of the business. A brief discussion on how these roles are addressed and implemented is discussed in Section 1.1.1. It is important to understand the value of leaders and managers in organization development and implementation because roles and responsibilities are also extremely important in facilitating a safe workplace for employees and productivity for stakeholders.

A safety and health program will not be ineffective without the engagement of leaders and managers setting the stage and serving as a model for others to follow. Leaders and managers are responsible for developing and maintaining an organization that facilitates safe performance of work. This task is not an easy task because there is always the human element that can add complexity from various avenues.

Workplace safety and health and productivity tend to flourish when:

- Management is intimately engaged
- Workers trust leadership and managers
- Workers are engaged

- The appropriate policies and procedures are in place and are followed
- Management set the stage by following the policies and procedures they put in place
- Culture is one that supports open and honest communication
- Workers have a feeling of belonging and are able to voice their opinions without fear of retaliation

Ideally managers are focused on implementing processes and systems in support of the goals of the organizations, and leaders are more focused on the people that are working within the system supporting those processes. In order for a safe and profitable workplace to become reality leaders and managers must be supportive and visible with their engagement.

The goal of a leader is to improve the process, system, or situation whereas the goal of a manager is for smooth operation of the processes, systems, or situations.[1]

1.1.1 Leadership and Management Role in Organization Development and Implementation

1.1.1.1 Responsibility for Organization Development and Maintenance

Leaders have the primary responsibility for organizational development activities with a high focus on the people aspects, and management has primary responsibility for operational and maintenance activities with a high focus on processes and systems. During organization development leaders are actively engaged in setting the vision and strategy development whereas managers are focused on ways to effectively implement the vision and strategy of the organization as set by leaders. Although management and leaders both have responsibility in all aspects of an organization (Figure 1.3), there are some situations where one or the other may take

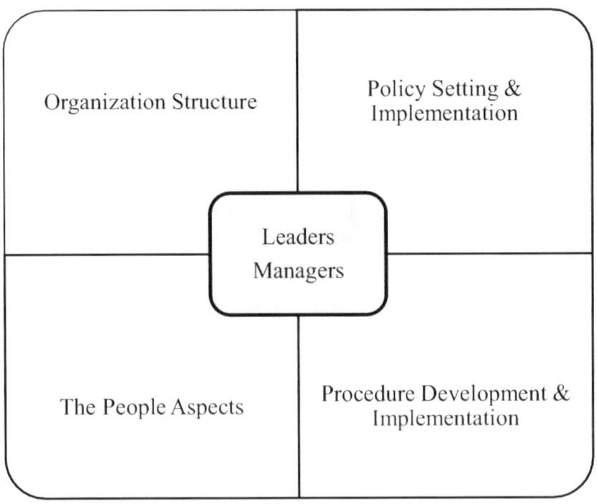

FIGURE 1.3 Managers and leaders' intersection of responsibilities

primary lead in accomplishing activities as it relates to achieving the vision and goals of the organization.

1.1.1.2 Organization Structure

Designing and putting in place an organizational structure that compliments and facilitates implementation of the vision and strategy of the business involves leaders that are visionaries and strategic. Once the structure is in place, management is instrumental in ensuring that the proper resources are in place to carry out and complement the organization's goals and facilitate safe work performance. Managers are engaged in obtaining the resources needed and implementation of the organization structure that is tailored to accommodate the needs of the business. This will involve retaining qualified health and safety professionals to ensure safe work planning and performance is kept as a focus area for getting work accomplished.

1.1.1.3 Policy Setting and Implementation

Leaders and managers should be actively engaged in policy setting. However, leaders may primarily be involved during organization design, and managers are primarily engaged during policy implementation. During policy setting and implementation it is pertinent that safety policies are developed and enacted with the express purpose of fostering a workplace that is focused on worker safety and health. It is also important for the leadership team to be first partakers and follow the policies that they develop and approve if they expect the rest of the organization to follow them. During this phase, safety policies are developed as an integral part of the cadre of policies being developed to dictate the operational parameters for the organization.

1.1.1.4 Procedure Development and Implementation

Procedures are developed to implement policies and strategies that have been developed to support business ventures. When developing procedures, ensure that they also provide the basis and instructions for hazard identification, analysis, and mitigation. In order for these procedures to be effective, they must be enforced and followed by managers and leaders alike. A focus on hazard prevention, mitigation, and control is at the forefront of employee safety while in the work environment and performing work activities in support of the organization goals and mission. Managers and leaders must keep in mind that procedure compliance is not just for workers, they themselves must comply with all procedures to set an example for the behaviors that they want others to emulate.

1.1.1.5 The People Aspects

Organizational leaders will often verbalize that people are their greatest assets. Developing and implementing a safety culture that facilitates a safe work environment that seeks to engage employees is a way to demonstrate the importance of workers. Workforce safety is a critical component of each element when designing and an organization to support implementation of the strategy for organizational success. These elements are multidirectional and can be modified in various orders at various times depending on the needs of the business. Modification often requires input from managers as they have primary responsibility for directing the day-to-day work activities.

1.2 COMPETENT SAFETY LEADERSHIP

Leadership is recognized as having a major influence on safety performance in organizations. Safety leadership can empower employees to actively engage in and assume responsibility for performing work safely. The goal of safety leadership (SL) is to improve safety performance (SP) within an organization. The Influence of SL on SP has been realized as a key factor contributing to the influences SP and SL can enhance employee participation in the work performed in organizations.[2]

Great safety leaders care deeply about people and spend the time needed to show them that they are valued through their actions. These leaders care about their employees as if they were a part of their family. When conducting business, they approach decision making with the knowledge that the life of a human being is the one thing that cannot be replaced. You can hire a replacement; however, the loss of an individual cannot be replaced. Embedded safety practices in organizations represent the fulfillment of the ethical and moral obligation to preserve human life.[3]

A safety leader is broadly defined as a leader that in their daily actions demonstrates safe behaviors in everything they do while encouraging and influencing others to do the same. Safety leadership refers to the interactions between leaders and followers, through which leaders can exert their influence on followers to exhibit behaviors that lead to a focus on safety and achieve safety goals. Effective safety leadership has been found to be a prerequisite for improving safety-related performance, through the motivation of workers to take ownership and responsibility for safety; reducing human failures associated with noncompliance; and fatigue and inadequate communication between leaders and workers. Safety leadership can have impacts on all aspects of safety performance that can be seen in the outcome of company metrics, the engagement of the workforce, and the number of workplace accidents and injuries. The impact of safety leadership is also visible in the behaviors exhibited by organization members and the way they approach and perform work.[4] Competent safety leadership can be found across the company in all ranks of an organization. A term that continues to gain popularity in the work environment is "safety leadership."

A type of leadership style that is favored by leaders in a work culture where employee engagement is desired as an integral part of facilitating work safely is transformation leadership.

Leaders that assume a transformational leadership style in support of safety demonstrate ten different actions that include:

- Expressing satisfaction when jobs are performed safely
- Rewarding achievement of safety targets
- Continuous encouragement for safe working
- Maintaining a safe working environment
- Suggesting new ways of working more safely
- Encouraging employees to openly discuss safety at work

- Talking about personal values and beliefs in the importance of safety
- Behaving in a way that demonstrates commitment to safety
- Spending time to demonstrate how to work safely
- Listening and responding to safety concerns

A transformational leadership style establishes the legitimacy and caring nature of the leader and their demands by appealing to the values of the individual.[5]

Competent safety leaders insert their influence through developing trusting relationships, demonstrating concern for employees and demonstrating their desire to operate in a regulatory compliant manner. Safety leadership must be visible among management/leadership, safety, and health professionals and employees who have been designated as workplace safety leaders. Safety leaders that are among the employee ranks are referred to as "workplace safety leaders." We will discuss these leaders and their contributions and importance in the upcoming section.

1.2.1 What Does It Look Like?

Safety leaders are not only found among the management rank but also among workers at all levels. Leaders at all levels through their interactions are the most effective way to influence others to become safety leaders. These leaders have the opportunity to influence their peers to change behaviors and embrace safety without realizing that change is occurring.

These leaders always follow company policies and procedures and demonstrate notable characteristics that can influence others such as:

- Encouraging workers to report all safety issues as they arise
- Take safety seriously as demonstrated by their words and actions and encourage others to do the same
- Institute and participate in measures to proactively prevent safety issues
- Trusted and respected by workers
- Have the ability to influence others at all levels of the organization
- The ability to inspire others at all levels of the organization
- Keen self-awareness
- Effective communication skills
- They never stop learning and encourage others to do the same
- They are respected by peers and management

Safety leaders spend a portion of their day walking the shop floor communicating with workers and peers seeking to understand the way work is being performed and looking for ways to improve work processes therefore improving safety. Leaders that have offices or work scope that does not permit them to be on the shop floor frequently should include in their calendars scheduled time to visit the shop flow and communicate with those performing tasks and seek to understand how work is performed and challenges that may be present.

Safety leaders that are among the ranks working with their peers provide one of the best ways to modify behaviors and get buy-in for facilitating a safe work environment and following procedures and practices that are geared toward safety while looking out for the safety of their peers.

1.2.1.1 Workplace Safety Leader

A workplace safety leader is an employee working with peers performing the same task as their peers. These individuals generally have experience in the work that they perform; therefore, they are familiar with and embrace safe work practices and follow procedures to ensure work is performed safely. These individuals are working alongside other workers while coaching them and influencing them in ways to perform work safely. A workplace safety leader is an individual who is trusted and respected, demonstrates a value for safety in the workplace, and models appropriate behavior for others to imitate. These leaders have charisma and the ability to influence others to follow their lead and actions. These leaders display many or all of the characteristics of a safety leader and are constantly being mentored and trained to continue to develop these characteristics and their influence among their peers and other organization members.

1.2.1.2 Managers That Are Safety Leaders

Managers that operate in the safety leadership realm are actively engaged in example setting and focus on ensuring the appropriate procedures and resources are readily assessed to facilitate safe work performance. These managers not only talk the talk, but they walk the talk. Meaning their actions and the words they speak are in alignment, and safety is truly a priority in the work environment. Management by walking around is embraced by these leaders because they understand that spending time with employees in the work environment offers the best advantage in understanding the way work is performed and modifying unwanted behaviors as they occur. It also represents an opportunity to get to know workers and understand their knowledge level and their capabilities.

1.2.1.3 Safety and Health Professionals as Safety Leaders

Safety and health professionals consist of a team of multidisciplinary professionals whose primary focus is to improve worker safety and health in the workplace. This team of professionals can include safety professionals, industrial hygienists, occupational health nurses, occupational medicine physicians, occupational health psychologists, ergonomists, occupational epidemiologists, and occupational toxicologists. The primary goal of this multidisciplinary team is to work in concert to design, implement, and maintain a comprehensive health and safety program that will enhance worker health and improve safety while supporting productivity.[6]

See Table 1.2 to review the role each team member plays in enhancing and preserving the health and safety of the workforce.

The fields of occupational health psychology and epidemiology have gained more importance in recent years, especially with the impact of COVID-19. The American Psychological Association (APA) lists the Coronavirus Pandemic as an epidemiological and psychological crisis. The APA cited reasons for the crisis stemming from the considerable number of people living in isolation, the number of job loss creating

TABLE 1.2
Safety and Health Professional Role in Worker Safety

Safety and Health Professional	Role in Preserving Worker Health
Safety professionals	Health safety professionals play a critical role in preventing injuries, illnesses, and death from occurring in the workplace. Responsibilities include assisting in procedural development to guide safe work performance, inspect and analyze the workplace conditions and practices to identify and prevent potential hazards, and assist managers in achieving and maintaining compliance with applicable health and safety standards
Industrial hygienists	An industrial hygienist is a professional whose primary responsibility is to protect workers from physical, chemical, ergonomic, and biological stressors. This is accomplished through anticipation, recognition, and control of hazards using procedures, regulatory requirements, air monitoring, and conducting health hazard risk assessments
Safety & health technicians	Work under the direction of the Industrial Hygienist primarily in the role of collecting air sampling and maintaining equipment
Occupational health nurses	An occupational health nurse assists in delivering programs and services that foster and maintain healthy and safe work environments. Responsibilities include assessing health risks, promoting wellness, preventing illness and injury through evaluation of workers and their work environmental potential health impacts, and implementing medical surveillance programs
Occupational medicine physicians	An occupational medicine physician specializes in the management of disabilities and the prevention of occupational illnesses, injuries, and diseases. Responsibilities include assessing and treating work-related injuries and illnesses; working closely with the industrial hygienist and safety professionals in identifying areas of risk and workers' health and safety potential impacts; and enhancing workers' health through clinical care, preventive medicine, and education
Occupational health psychologists	Occupational health psychologists' goal is to improve the quality of work life while protecting and promoting the safety, health, and well-being of workers in the workplace
Ergonomists	Ergonomists, sometimes referred to as human factors engineers, assist in the design of machines, tools, and other equipment to ensure that they can be used safely without inflicting harm on users. This is accomplished through data collection, accident investigations, work observations, and worker interviews in order to evaluate and propose changes to improve the ergonomic risk to workers
Occupational toxicologists	An occupational toxicologist is responsible for investigating the toxicity of chemicals found in the workplace and the adverse effects on workers and the environment
Occupational epidemiologist	Occupational epidemiologists conduct studies and evaluate worker exposure to a variety of chemical, biological, or physical agents to determine if the exposures to an agent result in the risk of negative health effects

a financial hardship for many, and the grief over the death of loved ones.[7] As many employers transition into returning workers to the workplace, there must be considerations given to providing the support workers will be needing in the workplace to deal with the anxiety, stress, and potential depression that they will be bringing to the workplace as a result of COVID-19.

1.2.2 Training Safety Leaders for the Present and the Future

Training is an important aspect that is used by industries across the globe to prepare workers to perform work safely and compliantly. One issue many face with training is that people tend to lose knowledge over time, especially if the knowledge is not used frequently in performing their work activities. Another issue is the introduction of many training topics in a brief period of time at the onset of employment. Employees tend to walk away from receiving an onslaught of training confused with many questions that often go unasked and therefore unanswered. A good practice would be to administer training over a period of time. In such a case a new employee should be paired up with an employee that is recognized as a workplace safety leader to guide new employees on how to perform work safely while taking into consideration themselves and their coworkers.

The role that safety leaders play in the success of companies in terms of implementing safety programs and facilitating safe performance of work makes them invaluable resources. As such, management should invest in training employees to take on the role of a safety leader. Training leaders to deal with safety issues today and in the future can be cost-effective and distinguish a marginal company from a good company. The cost of dealing with safety incidents and employees that have suffered an injury or illness can be astronomical. The National Safety Council (NSC) estimated for 2021 a total of $167.0 billion dollars was spent on injuries that occurred in the workplace. The cost of $167.0 billion includes productivity and wage lost in the amount of $47.4 billion; medical expenses of $36.6 billion; administrative fees of $57.5 billion; employers' costs of $13.8 billion; damages to motor vehicles in work-related injuries of $5.4 billion; and fire losses of $6.3 billion.[8] This estimate of 167.0 billion includes income not received or the expense incurred because of fatal and nonfatal injuries that were preventable. The cost estimates for 2021 provide context into the economic impact of preventable injuries. Some of the costs of dealing with safety incidents are visible, and other costs are not always seen in the near term. Some hidden (invisible) costs include:

- Loss of worker morale
- Loss of wages
- Paperwork process
- Worker distraction
- Replacement worker training cost
- Legal costs
- Injured person time away from work
- Supervision lost time

- Time lost to replace damaged equipment
- Workers' compensation premium (increase)
- Medical expenses
- Increased overhead cost
- Loss of efficiency resulting from inserting new employees with less experience

The list presented only represents some of the losses that can result from a safety incident. Another particularly important loss that is not included is the potential loss of company reputation. Some may view the company as an unsafe place to work, and such a reputation may have an impact on their ability to attract capable staff.

Safety leadership training can take on many forms (Table 1.3) from on the job to classroom. Regardless of the form, learning can take place as long as the information

TABLE 1.3
Safety Leader Training Delivery Methods

Delivery Method	Potential Benefits
Classroom	Training conducted in a classroom environment having an instructor delivering the topics and guiding discussions and learning materials introduced
On-the-job training	Providing hands-on training on conducting actual workplace tasks correctly and safely
Mentoring	Sharing knowledge on how to perform work safely
Coaching	Providing guidance to workers on the proper way to perform work activities safely through observation and instructing activities
Case studies	Developing critical thinking, problem solving, and analytical skills. The scenarios can be real or imaginary, but in the context of employee training, they all illustrate situations at work
eLearning	Using digital means such as cell phones and computers to deliver training topics provides flexibility in training and learning
Video-based learning	Learning techniques can be done in groups or individually. Involve in watching safety-related videos containing various topics and safe work practices. When performed in a group, the videos can be used to spark discussion on in-house applications
Roleplaying	Role playing involves addressing safety-related issues or providing explanation or information on these issues allowing the trainer and the trainee to place themselves in each other's role
Board messaging	Visual safety message placed strategically in the work area in places of high traffic (for example, lunchrooms and break areas)
Required reading	Publish and distribute safety information that every employee is required to review to include management
Audio messaging	Safety message introduced daily at the beginning of the shift by a member of the leadership team
Story telling	By employees highlighting safety-related incidents that happened to them along with their recovery method and lessons learned

TABLE 1.4
Safety Trained Leader Training Courses

Potential Course	Course Content and Purpose
General safety	To expose workers to the various safety issues that can be present in the workplace and measures that can be taken to enhance safe work performance
Personal protective equipment (PPE)	Knowledge of the types of personal protective equipment used in the work and when to use each PPE as well as their limitations
Communication	Training prepares safety trained leaders to communicate with a diverse population of workers while increasing the understanding and knowledge of the different ways people communicate
Listening skills	Training in attentive listening enhances the quality of communication and can aid safety leaders in their ability to influence others to embrace safety in the workplace
Culture	Training on cultural attributes such as communication, engagement, trust, following procedures, looking out for the safety of coworkers. These elements are key in shaping a safe inclusive work environment
Ergonomics	Training that covers principles to help workers perform physical tasks with more comfort and less stress on your body and knowledge to set up their workstation in a safe configuration
Hazard evaluation	Training on how to evaluate work tasks, identify potential hazards, and determine potential mitigation measures
Hazard communication	Hazard communication training provides employees with knowledge of the chemical that they can potentially be exposed to in the work environment and protective measures to avoid exposures
Safe work practices	Training on how to perform work safely and some of the applicable regulatory requirements associated with activities performed on a consistent basis
Lifting techniques	Prescribe the proper techniques to use when lifting materials to avoid inadvertent injury
Emergency response	Training on the types of incidents that may occur in the work environment and the appropriate response actions to take to protect workers and property
Signs and barricade	Training on recognizing the various signs in the work and barricades used and having an understanding of the associated work actions to increase safety
Behavior-based safety	Training that can influence employees to consider safety with each action taken seeking a safer outcome and preventing incidents

is delivered by someone who has the knowledge and respect that compliments the material delivered. Before implementing training, delivery method care must be given to ensure that the appropriate learning environment has been selected for delivery. Table 1.4 contains topics that should be considered for inclusion in the company training plan to enhance the success of training safety leaders. This table only offers some course suggestions as there are other courses that may be selected based on the company goals, objectives, and regulatory compliance posture.

REFERENCES

1. Booher, D. *Communicate Like a Leader: Connecting Strategically to Coach, Inspire, and Get Things Done.* Berrett-Koehler Publishers Incorporated, 2017.
2. Zhao, Linyi, Yang, Daojian, Liu, Suxia, Nikrumah, and Kwame, Edmund Nana. The effect of safety leadership on safety participation of employees: A meta-analysis. *Frontiers in Psychology*, 13, 2022, https://doi.org/10.3389/fpsyg.2022.827694, June 16, 2022.
3. Safety & Health Magazine, Safety Leadership: Seven practices of great safety leaders, https://www.safetyandhealthmagazine.com/articles/19549-safety-leadership-seven-practices-of-great-safety-leaders, March 29, 2020.
4. Ta, My Thi Diem, Kim, Tae-eun, Gausdal, and Haugen, Anne. Leadership styles and safety performance in high-risk industries: A systematic review. *Safety and Reliability*, 41(1), 10–44, 2022, https://doi.org/10.1080/09617353.2022.2035627
5. Pilbeam, Colin, Doherty, Noeleen, Davidson, Ross, and Denyer, David. Safety leadership practices for organizational safety compliance: Developing a research agenda from a review of the literature. *Safety Science*, 86, 110–121, 2016.
6. United States Department of Labor, Occupational Safety and Health Administration, Occupational Health Professionals, https://www.osha.gov/occupational-health-professionals
7. Psychological Impact of COVID-19, https://www.apa.org/topics/covid-19/psychological-impact,American Psychological Association, April 16, 2020
8. https://injuryfacts.nsc.org/work/costs/work-injury-costs/

2 Leadership Selection and Development

2.1 SELECTING AND DEVELOPING LEADERSHIP FROM WITHIN

Selecting the right leaders at the right time for the right role to fill current and future needs is not a simple yet an ongoing task. It is even more complicated in today's world to select the right leader as there are so many different available opportunities with a diverse mixture of organizations offering a wide array of opportunities, benefits, and perks. Organizations must be able to compete to attract and retain the best leaders to support their needs. There may be times when attracting personnel may not be so challenging; however, retention is difficult. Many organizations have turned to seeking and selecting leaders from the ranks of those that have been working within the company. The benefit of this strategy includes internal candidates who already have an understanding of the goals of the company and the culture that is embedded in the way business is conducted. Leaders that have been grown and nurtured internally may have an advantage over external hires in that they should have a keen understanding of the internal culture, procedures, policies, and acceptable practices. In addition, the visibility of promotion from within may provide incentives for others to remain with the company and become more committed to the goals and work to seek opportunities for advancement within the company.

Just as there are benefits of elevating from within, there are some challenges as well. One important benefit that may be missed is the opportunity to take advantage of the different ideas and ways of conducting business. With internal candidates, there is a high potential that business will be conducted as usual, and there can be a lack of infusion of innovative ways of completing work and business may remain the same or become stagnant. Even with this challenge, the benefits of hiring from within is an effective way to go for most organizations for some of their leadership positions. In order to potentially overcome this challenge, it is necessary to take advantage of providing mentoring or training that will tap into the new leader's ability to infuse creativity into business decisions and practices as well as their ability to think and act strategically.

Promoting from within is a great tool to leverage internal candidates for key positions within an organization. The message this action sends to the employees is that the company is willing to invest in them and that they are considered valuable resources. Some reasons that an organization may choose to promote from within include:

- Promoting from within can increase employee loyalty to the organization
- Promoting from within can serve as a retention tool to retain critical staff

DOI: 10.1201/9781032624389-3

- Promoting from within can produce multiple opportunities for employees because the
- individual being promoted will need to be replaced
- Promoting from within can serve as a motivator for other workers to work harder and strategically toward promotion
- Promotion from within is generally less costly than perusing external hires
- Inside hires are already knowledgeable and acclimated with the culture, policies, procedures, and practices of the organization
- Inside hires already have relationships with some employees and should become productive in their new roles sooner than an external hire.[1]

The option of developing leaders internally in preparation to take on senior roles within the company is a good option and can present a "win-win" situation for the company and the employees.

2.1.1 How to Grow Leaders from Within

Growing leaders from within must take on a strategic and targeted approach if success is to be achieved. The approach taken must be in line with and complement the business strategy and vision of the organization. The process of raising leaders from within is time-consuming and represents a great investment for the company. To accomplish this task there must be a commitment of time and money for success to be achieved, yielding leaders ready to take on new roles and move the company into the future. A good approach to developing people begins with a succession plan designed to address strategic positions and not all positions at once.

Organization leaders have come to the recognition that the next generation of leaders may already be a part of the department or working elsewhere in the company. Succession planning has become an attractive option for identifying and developing leaders to replace those that are retiring or leaving in those organizations that are not challenged by reduced resources, low headcount, and rapid growth. Succession planning is used to increase the availability of internal candidates to fill key positions as they become available. When used, succession planning is viewed as an important part of a company's business strategy. There are many changes within a business process that can facilitate the need for consideration of developing and implementing a succession strategy. These activities may include

- Changes in the corporate business strategy
- Changes in the use of technology
- Knowledge retention needs
- High-employee turnover
- Implementation of process changes
- Organization restructuring
- Development and implementation of a retention strategy

The inability to implement an effective succession plan can result from the following:

- The inability to maintain skilled staffing eligible for succession
- The plan is too aggressive and not easily implementable
- The strategy is not communicated to the workforce
- Lack of leadership commitment and demonstrated support
- Lack of buy-in from employees

2.1.2 Developing Leadership Skills

Some leaders are of the school of thought that leadership is a natural skill that one is born with. However, many schools of thought have supported the premise that leaders can be trained to become and function as great leaders when being subjected to the appropriate training model. Leadership development is an act of capacity building for the expressed purpose of nurturing leaders in their ability to anticipate various challenges when working through complex issues. The development of leaders should be introduced in a structured progressive manner that moves them from novice to intermediate then to a level of having the skills to deal with extraordinarily complex issues.[2] Having knowledge of the characteristics good leaders possess, the appropriate training program can be developed and instituted to develop great leaders. Characteristics of good leaders include the following:

- They are strategic and purposeful in action
- A good relationship builder
- A good listener
- Deal well with uncertainty
- Keep commitments
- Actively engaged in the business
- Skilled communicator
- Exhibit trustworthy behaviors
- Willing to delegate
- Assume accountability for their roles and actions
- Able to build and retain trust
- Concern for the safety and well-being of employees

Developing leaders and nurturing the skills that they already possess will move them into becoming leaders for today and tomorrow can be accomplished through various avenues. These avenues include:

- Classroom training
- Coaching
- Mentoring
- Reading (books, articles)

- Shadowing – allow perspective leaders to shadow other leaders on projects (participating in key discussions, planning, and decisions) that are critical to the role they are preparing to assume
- Job rotation

Training development methods can be used at various stages throughout the developmental process in any order deemed necessary to achieve optimal results. In addition, it may not be necessary to take advantage of all training methods to provide the training necessary for leaders to become and remain effective in their roles.

2.1.3 First a Follower and Then a Leader

Leaders have always been viewed as the most important people contributing to the success of the business. It is believed that leaders contribute approximately 20% to the success of an organization whereas followers are among the remaining 80%. Most people follow more than they lead in a lifetime and leaders cannot lead without having followers among the ranks. In fact, it is feasible for an employee to assume the role of a leader and a follower interchangeably.[3] For example, in organizations, leaders have a reporting chain of command that they follow, and the president of an organization in most chains have reporting responsibility to a Chief Financial Officer (CFO) or a board of directors where they are required to follow directions from the board.

A follower is a person who actively seeks to pursue the actions as directed by the leader to ensure that the goals of the organization become a reality. Effective followers recognize the authority of the leader and make a conscious decision to follow for the good of the project or enterprise. Followership is not about doing as you are told, but it is built upon the premise of the follower's belief in the organization and its mission, willingness to work toward the greater good of the organization, and a sense of loyalty. Leaders must have the ability to influence and inspire others to follow. This is essential in a work environment where employee safety is challenging and a core value of the company.

Followers can fall into various categories or types. To demonstrate the challenges leaders may face when engaging with followers the five styles of followership found in Table 2.1 are introduced.

Table 2.2 points out some of the challenges that can be encountered when dealing with the different types of followers and the benefits or lack thereof that some types of followers may bring to a team. The preferred type of follower that can provide more value to a leader is one that exhibits the attributes of a Star Follower. This type of follower can add values such as assisting the leader in devising the best decisions or solutions through challenging the leader through their willingness and ability to question decisions and information provided by the leader. In addition, a Star Follower can serve as an example to others in the organization and influence them to adopt like behaviors. A leader will receive complete support from a Star Follower when they agree with the leader. However, in the event of disagreement the

TABLE 2.1
Followership Types and Description

Followership Styles	Description
The Sheep	This style when used seeks for the leader to do all of the thinking and motivate the follower to perform
The Yes Person	These followers are always positive and are always on the side of the leader yet looking for the leader to do the thinking for them
The Alienated	These followers think for themselves, however, exude a lot of negative energy. They are cynical and skeptical of every decision or action
The Pragmatic	Sits on the fence and goes with the direction of the wind
The Star Follower	Thinks for themselves and actively exuding positive energy[4]

TABLE 2.2
Followership Types and Potential Ramifications

Followership Styles	Potential Ramifications or Pitfalls
The Sheep	This type of followership style when used does not seemingly add value to a discussion or development of solutions since the follower needs to be motivated and is seeking to have the leader do the thinking for them. The leader does not have the benefit of someone challenging their thoughts and assumptions that can lead to the best solution to a problem having input from others
The Yes Person	These followers having positive attitudes can have positive impacts on other team members. However, these followers may not add objectivity to discussions or decisions as they are always on the side of the leader not questioning decisions or actions while depending on the leader to think for them
The Alienated	Although self-thinkers the negative energy exuded can impact negatively on team morale and performance. The Alienated can be cynical and skeptical of every decision or action
The Pragmatic	This type of follower adds essentially no value to team's activities or developing solutions because of their indecisive ability
The Star Follower	This type of follower possesses the best characteristics to support a leader and the goals of the organization because as active thinkers they can contribute to problem solving and have the ability to introduce solutions that the leader may not have considered, and the positive energy they exude can have a positive impact on teammates and productivity

Star Followers will offer constructive criticism and alternatives that can be of help to the leader.

There are many benefits realized by an organization when employees trust their leaders. These benefits include the following:

- Trusted leaders empower employees to think and make thoughtful decisions
- Trusted leaders facilitate productivity
- Trusted leaders inspire workers to create and produce without the fear of potential failure
- Trusted leaders encourage focus on the task to be accomplished
- Trusted leaders foster innovation and novel ways of solving problems
- Trusted leaders can be instrumental in facilitating an employee's desire to remain with the organization[5]

2.1.4 Selecting Leaders

Creating an environment that is one of collaboration, authentic, personalized, and innovative requires leaders that exhibit behaviors that reinforce the attributes of the desired environment. The leadership behavior that can be successful in creating the aforementioned environment includes a collaborative mindset (inclusivity decision maker and solicit feedback from team members), developer of people (involve in mentoring and coaching staff and provide honest feedback), digital technology savvy (strategic use of technology when communicating), and a global citizen (exhibit a diverse mindset) and has the ability to anticipate and plan for the future.[6]

It has been established that leaders play important roles in organization. As such, it is critical that these roles are filled with people that can add strategically to all aspects of an organization. Selecting leaders should be done with the needs of the organization in mind to address current and future needs. Before developing a job posting, invest the time needed to evaluate the organization and develop the role and responsibilities for the leader being sought.

Figure 2.1 lists important attributes and examples of each attribute that should be considered when seeking leaders:

- What characteristics are needed in the leader?
- What are the issues in the organization that need addressing?
- What organization function needs a leader?

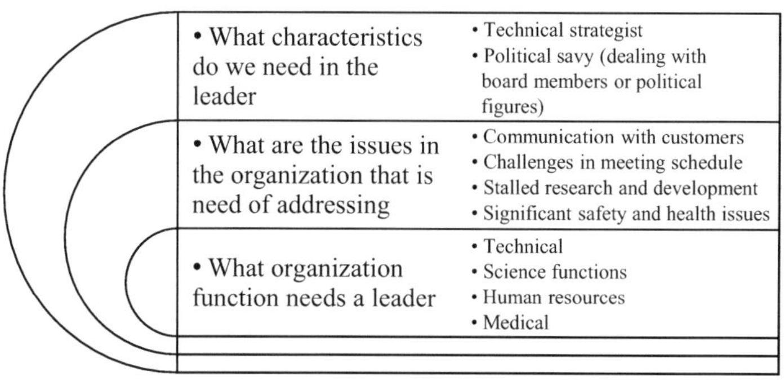

FIGURE 2.1 Leader selection considerations

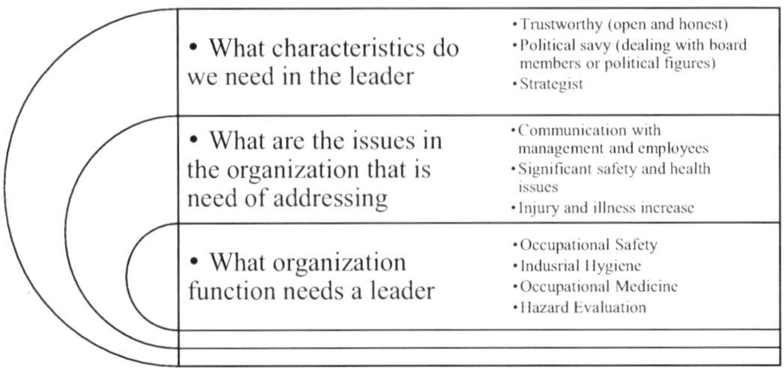

FIGURE 2.2 Safety and health leader selection considerations

Selecting leaders is highly dependent upon the attributes listed in Figure 2.1 at a minimum. Other attributes may be just as important depending on the position being filled and the business strategy. For example, if seeking a leader for a technical function depending on the function there may be a requirement for the leader to possess some technical knowledge in the area of interest.

The attributes listed in Figure 2.1 are examples and may not apply to the needs of all organizations. It is presented to show the potential thought process one should engage in when selecting leaders. Let's consider the need to select a safety and health leader to direct and lead a staff of safety and health professionals. The manager/leader should be expected to provide the leadership, vision, and resources needed to implement an effective safety and health program. In doing so it is expected that the leader would:

- Facilitate a culture that makes worker safety and health a core organizational value.
- Demonstrate a commitment to eliminating hazards, protecting workers, and continuously improving workplace safety and health within the organization.
- Provide the needed resources to develop, implement, and maintain the safety and health program.
- Visibly demonstrate and communicate their safety and health commitment to workers, leadership team, stakeholders, and others.
- Set an example through their own actions and interactions with others.

Using the format listed in Figure 2.1 and tailoring it to the selection of a Safety and Health Manager the process may resemble the selection considerations listed in Figure 2.2.

2.2 SELECTING AND HIRING LEADERS EXTERNAL TO THE COMPANY

Identifying and hiring leaders externally to a company is an investment that takes time and resources (money, people). The process typically begins with a job posting and/or selecting an agency to develop a list of candidates, conduct the initial screening, and provide a limited list for the organization to conduct additional interviews. When hiring externally, the interview process focuses strongly on the person-organization fit and use of the behavior-based interview techniques. These techniques do not always produce the optimal candidates for the role, but when used properly, they provide a screening mechanism that provides the opportunity of selecting the right candidate for the job.

The Person-Organization Fit Assessment (P–O Fit) is designed to assess the compatibility between an individual, the organization, and culture of the hiring organization.[7] Theoretically, just because a leader is recognized as being a good leader in previous roles does not mean that the leader will be a fit for every organization and culture. Behavior-based interviews can be used to potentially predict a leader's future performance by assessing their past performance. The interviews and questions asked are customized to the specific organization and position. This customization of interview questions is used to determine how the applicant performed in specific employment situations during past roles and the resultant outcome from their behavior. For example, in an organization where employee safety and health is declining and of issue, questions would be structured to assess the candidate's position on developing and maintaining a safe work culture. Behavior-based interviews are most effective when conducted in panel format having two to four interviewers. The interview questions should be designed to guide the interviewee to discuss the situation in which they demonstrated competencies of interest, the actions they took, and the result of their actions.

2.3 LEADERSHIP ETHICS: WHY IS IT IMPORTANT?

Leadership behavior is more than allocating resources, directing staff, or building the organization as a cohesive team. True leadership is engaged in moving followers to embrace and support the vision that the leaders have developed to ensure that the mission of the organization becomes a reality. Studies have shown that it is not enough to put in place leaders that are competent in technical matters, intelligent, and industrious as these managers may be ineffective if they are arrogant, untrustworthy, abrasive, self-absorbed, and compulsive.[8] The totality of the effectiveness of a leader does not stop at the role they play in organizational matters, but it encompasses the behaviors that they exude in accomplishing the activities assigned to their role. There is a growing realization that ethical leadership is important and plays a great role in a leader's ability to effectively lead, gain and retain respect, and tap into critical knowledge possessed by followers.[9]

Ethical leadership has been a continuous topic of interest for scholars and practitioners as researchers have found that there are several positive outcomes related to ethical leadership within organizations. Among these positive outcomes are:

- Better employee work performance
- Positive organizational citizenship behaviors (employees willing to go beyond what is required)

- Increased employee commitment
- Increased employee job satisfaction

Just as ethical leaders can have a positive influence on organizations, unethical leaders can have negative influence on an organization such as employees may be more likely to adopt the unethical practices of the leader.[10] Concerns over unethical leadership behavior continue to persist in many workplaces. After learning of a leader's unethical activity some employees may choose to continue to provide support whereas others will withdraw their support for the leader. Unethical behavior is any behavior that goes against the accepted moral norm. When an employee observes leaders acting in an unethical manner, their attitudes toward that leader can change and become one of negativity. In addition, employees tend to lose trust in the leader. Some unethical acts committed by leaders include:

- Falsifying business expenses
- Making racially insensitive jokes
- Verbally abusing employees
- Taking credit for subordinate ideas or inventions
- Harassing or bullying employees[11]
- Lying
- Theft of company properties
- Nepotism
- Favoritism
- Requiring employees to work during non-work hours without pay

A study conducted by Asnakew and Yibeltal in 2019[12] using data from 262 employees of various organizations in Ethiopia confirmed the detriment of unethical leadership. The results of the study showed that unethical leadership behavior has a significant effect on followers' departing from usual or accepted standards in the workplace brought about as a result of their perception of organizational politics and the lack of fairness. Although the study took place in Ethiopia, one can surmise that the results can be applicable to other countries and workplaces to a relative degree. It is believed that unethical leadership can negatively affect individuals as well as organizations. Unethical behavior of leaders and managers can facilitate followers developing feelings of anxiety, helplessness, frustration, job dissatisfaction, and loss of trust.

Ethical leaders make ethical decisions when dealing with company business and the workforce. These leaders have the capability to assist the company in building their reputation as being preferred clients and gain and retain the respect of others in their industry. Examples are provided below to answer the question directly as to why leadership ethics are important considerations that must be given to the leader as well as the company. Over the past decade many companies have had senior leaders exposed as a result of their unethical behavior while conducting company business. Examples of some of these exposures include:

- Jeffrey Skilling – a former chief executive of Enron whose lies contributed to the collapse of Enron. The fall of Enron cost shareholders billions of dollars and employees their retirement savings. Mr. Skilling was convicted of 18 counts of fraud and conspiracy along with one count of insider trading for his in the accounting schemes that concealed Enron's debts and weak finances from shareholders and regulators. Mr. Skilling was initially sentenced to more than 24 years in prison; however, that term was reduced later to 14 years.[13]
- Tyco International – Former Tyco International CEO Dennis Kozlowski and a subordinate were convicted of looting more than $600 million from Tyco to pay for lavish parties, fancy art, and an extravagant Manhattan apartment. Kozlowski and Swartz were accused of enriching themselves by taking unauthorized pay and bonuses, abusing loan programs, and selling their company stock at inflated prices after being untruthful about the company's finances.[14]
- Philip Clarke – A former chief executive of Tesco stated that under the leadership of Clarke, the company "allowed the trust of its millions of customers to be 'eroded', that it lost its reputation for having low prices and that there was a 'failure' of leadership." The failure of leadership included eliminating too many senior executives that impacted the company's culture and fall out between the company and one of their biggest suppliers.[15]
- John Stumpf – former Wells Fargo executive, engaged in a deceptive practice of opening accounts for customers without their knowledge. The misconduct affected millions of bank customers from 2002 to 2016. Banking regulators described Mr. Stumpf's actions as a "systemic" misconduct in sales practices. As a result of his actions, he was barred from the banking industry and fined $17.5 million.[16]
- Scott Thompson – Chief executive officer at *Yahoo! Inc.* resigned after it was revealed that his academic credentials were inflated. A principal shareholder discovered that Thompson's official Yahoo biography exaggerated his academic credentials, taking credit for having a dual degree in accounting and computer science, whereas his actual degree awarded was solely in accounting. Thompson's departure was costly for Yahoo as he was paid an annual base salary of $1 million, 1.5 million in cash bonus, and millions in stocks and other perks.[17]

The above cases represent a small segment of unethical behaviors perpetrated by corporate leaders over decades. This trend of unethical behavior is destined to continue and reported be at some frequency as they are discovered. The impact of these behaviors has negatively impacted the companies, stakeholders, and employees.

Leadership ethics can impact a company in various ways from contributing to the loss of key talent, loss of company reputation, tarnished reputation, or financial viability. Ethical business practice is a concern for companies as such these companies offer annual training to remind workers of their expectations for conducting business in an ethical manner. Although ethics training is mandated, and many complete the training, unethical business practices continue and are consistently called out, and these leaders are exposed. However, they must continue to be a focus on challenging leaders to adopt ethical practices when conducting business. A safety leader must act in an ethical manner in order to gain the trust and respect of workers and encourage them to adopt safe work practices.

REFERENCES

1. Alston, Frances. *Lean Implication Applications and Hidden Costs*, CRC Press, 2017.
2. Khoshhal, K.I., and Guraya, S.Y. Leaders produce leaders and managers produce followers. A systematic review of the desired competencies and standard settings for physicians' leadership. *Saudi Medical Journal*, 37(10), 1061–1067, 2016. https://doi.org/10.15537/smj.2016.10.15620
3. Schindler, James, and Schindler, James. *Followership: What It Takes to Lead*, Business Expert Press, 2014. *ProQuest Ebook Central*,
4. *The Art of Followership: How Great Followers Create Great Leaders and Organizations*, edited by Ronald E. Riggio, et al., John Wiley & Sons, Incorporated, 2008. *ProQuest Ebook Central*, https://ebookcentral.proquest.com/lib/osu/detail.action?docID=331447
5. Alston, Frances. *Culture and Trust in Technology-Driven Organizations*, Taylor & Francis Group, 2013. *ProQuest Ebook Central*, https://ebookcentral.proquest.com/lib/osu/detail.action?docID=1408091
6. Meister, J. C., and Willyerd, K. *The 2020 Workplace: How Innovative Companies Attract, Develop, and Keep Tomorrow's Employees Today*, Harper Business, 2010.
7. McEntire, L. E., and Greene-Shortridge, T. M. Recruiting and selecting leaders for innovation: How to find the right leader. *Advances in Developing Human Resources*, 13(3), 266–278, 2011, https://doi-org.oregonstate.idm.oclc.org/10.1177/1523422311424712
8. Mendonca, Manuel, and Kanungo, Rabindra. *Ethical Leadership*, McGraw-Hill Education, 2006. *ProQuest Ebook Central*, https://ebookcentral.proquest.com/lib/osu/detail.action?docID=316303
9. Babalola, M. T., Stouten, J., Camps, J., and Euwema, M. When do ethical leaders become less effective? The moderating role of perceived leader ethical conviction on employee discretionary reactions to ethical leadership. *Journal of Business Ethics*, 154(1), 85–102, 2019. http://www.jstor.org/stable/45022861
10. Sam, C. H. What are the practices of unethical leaders? Exploring how teachers experience the "dark side" of administrative leadership. *Educational Management Administration & Leadership*, 49(2), 303–320, 2021. https://doi-org.oregonstate.idm.oclc.org/10.1177/1741143219898480
11. Fehr, R., Fulmer, A., and Keng-Highberger, F.T. How do employees react to leaders' unethical behavior? The role of moral disengagement. *Personnel Psychology*, 73, 73–93, 2020, https://doi-org.oregonstate.idm.oclc.org/10.1111/peps.12366
12. Asnakew, Zeleke, and Mekonnen, Yibeltal. Unethical leadership and followers' deviance: The mediating role of perception of politics and injustice. *The Journal of Values-Based Leadership*, 12(1), Article 12, 2019, https://doi.org/10.22543/0733.121.1250
13. Skilling, Jeffrey. Former Enron Chief, Released After 12 Years in Prison, February 22, 2019, https://www.nytimes.com/2019/02/22/business/enron-ceo-skilling-scandal.html
14. NBC News, Ex-Tyco CEO Dennis Kozlowski found guilty, June 17, 2005, https://www.nbcnews.com/id/wbna8258729
15. BBC News, Tesco: Where it went wrong, January 19, 2015, https://www. bbc.com/news/business-30886632

16. The Washington Post, Former Wells Fargo chief executive barred from baking industry, January 23, 2020, https://www.washingtonpost.com/business/2020/01/23/banned-life-former-wells-fargo-chief-executive-barred-banking-industry/
17. Yahoo News, Yahoo CEO Thompson Resigns, May 15, 2012, https://news.yahoo.com/news/yahoo-ceo-thompson-resigns-1605200001.html

3 Determining Regulatory Applicability

3.1 EVALUATING THE BUSINESS RISK PROFILE

A top priority for business owners and leaders is to understand the risk profile of the business. In order to gain an understanding a comprehensive evaluation must be conducted by qualified knowledgeable individuals to produce a baseline that can be used by leadership to enhance compliance and sustainability. When speaking of a business, risk profile reference is made to the willingness and ability of leaders of a given business to take risks while taking into consideration the threats to which an organization is exposed. Organizations use a risk profile when determining investment allocation for their portfolio and as a way to mitigate potential risks and foreseeable threats. An effective risk profile includes considerations for activities such as the cost associated with the projected risks, the nature of the threats that a business can face while meeting their business strategy, the degree to which the threat can adversely impact the business, and the controls that an organization should put in place to mitigate or manage the risk. These are just some of the considerations that must be addressed when developing a risk profile for any given business. Included in a comprehensive risk profile and registers are considerations for identification and methods for complying with the many laws and regulations that are applicable to the respective business/organization and the health and safety of employees.

Risk management is a topic of concern for many corporate executives since there are some levels of risk associated with all business ventures. When launching a business venture there are many aspects of the business strategy that must be considered to ensure the appropriate level of risks are accounted for to launch a sustainable business. Included in the strategy are the various laws and regulations that must be evaluated for business applicability. In order to accomplish this in a manner that provides the best compliance profile for a business, there needs to be a clear and comprehensive business plan and strategy of what the business role in society looks like such as products produced and services that will be provided. Therefore, the leadership team must be cognizant of the various rules and regulations that will apply to their business activities, include these laws in their operational procedures, and have a knowledgeable staff to assist with program development and implementation. The lack of implementation of an applicable law can cost a company in various ways such as fees and penalties, loss of reputation, frequent audits, lawsuits, or even the smooth running of health and safety programs.

Regulatory compliance occurs when a business follows state, federal, and international laws (if conducting business internationally) and regulations that are applicable to their business. These laws and regulations have been developed and evolved

DOI: 10.1201/9781032624389-4

over time based on specific needs and past practices. For example, during the late 1960s occupational injuries and illnesses were increasing in number and severity. Disabling injuries increased by 20% during the decade, and approximately 14,000 workers were dying on the job each year. This degradation in workplace safety prompted the necessity for prompt action. Finally on December 29, 1970, President Richard M. Nixon signed The Occupational Safety and Health Act of 1970, which is also known as the Williams-Steiger Act in honor of the two men who pressed so hard for its passage. The Occupational Health and Safety Act (OSHA) was originally created because of the public outcry against rising injury and death rates experienced by workers in the workplace. Today OSHA continues its focus on reducing injuries, illnesses, and fatalities in the workplace while considering future challenges such as new chemicals and other hazards in the workplace, growing service sector industries, and changing workforce needs. After many decades of existence OSHA continues to be a driving force in facilitating workplace safety.[1]

Health and safety laws and regulations are designed to provide protection for workers and the environment as well as reduce liability risk for a company as such; it is paramount that a company evaluate the business and the many different regulations that they must build their business strategy around to achieve and maintain compliance. This is easier said than done because of the large number of regulatory requirements that change from time to time. Addressing risk is further complicated depending on business with diverse and complex strategies that will undoubtedly require adherence to a large cadre of regulatory requirements.

In this chapter we will discuss considerations for determining the applicability of the various laws, developing programs to ensure regulatory requirements are embedded in the business strategy and the way business is conducted, and developing and performing periodic assessments to ensure compliance is maintained. In addition, many laws will be introduced and discussed to provide context of the importance of regulatory compliance. Although many laws will be introduced, not all of the laws that can assist in protecting workers, the environment, and companies as leaders attempt to complete their business strategies will be discussed. An important aspect to remember is that when evaluating applicability of the many laws to a business venture, a holistic approach must be taken to ensure all applicable laws are included in the implementation strategy. In addition, it is important to understand the industry in which you are competing as some industries have what is known as a strong oversight compliance culture.

3.1.1 Employment Laws Facilitation of Worker Safety and Health

There are a host of laws and regulations that have been instituted to guide workers and companies in areas such as on fair treatment of employees, safety and health, and conducting business ethically. These laws are placed in the category of employment laws because they address the rights and considerations for the individual employee and their work environment. These laws are legally binding, and an employer is expected to comply with all applicable laws and regulations. A breach of any one legally binding law can lead to penalties for the company and in some cases the prosecution of senior leaders.

There are certain laws and regulations that all businesses must be compliant with regardless of the industry or company venture. As such conducting an evaluation of applicability is not as important as with other laws. These laws can be placed in several categories and are listed and briefly discussed on the Department of Labor web page.[2] Although targeted toward a specific topic or issue, these laws also play a role in ensuring the health and safety of workers. In some cases, directly and indirectly, as dictated by laws in the following categories:

- Employment termination
- Discrimination, retaliation, and harassment
- Worker's compensation
- Family medical leave
- Fair wage (minimum wage, overtime pay)
- Safe workplace laws

One important aspect of a safe workplace is having employees that are focused on accomplishing work tasks safely and adhering to the company's safety and health policies, procedures, and practices. Employees can be distracted from accomplishing work if they are concerned with issues such as their ability to take time off to recover from injuries and illnesses, attend to family emergencies, dealing with workplace harassment, or even concern about their pay. These employment laws provide an avenue to address these issues so that a worker can feel comfortable and at ease focusing on the work activity that they were hired to perform. These laws can be viewed as directly or indirectly impacting safety and health in the workplace depending on individual points of view and perspectives.

TABLE 3.1
Workplace Laws Supporting Directly or Indirectly Safety and Health

Laws	Potential Safety and Health Implications
Employment Termination	Eliminates or reduces the stress associated with unfair job loss
Discrimination and Harassment	Reduces distractions and stress associated with unfair treatment in the workplace
Worker's Compensation	Reduces the stress associated with on-the-job injuries and illnesses. Aids employees that are injured in getting the health care needed and remain away from the workplace until they are released for work by a physician
Family Medical Leave	Reduces injuries that can occur as a result of workers being in the work area while focusing on family issues
Fair Wage	Supports workers' ability to focus on work and perform work safely. Aids in the elimination of distractions
Safe workplace Laws	Specifically targets workplace and worker safety

The laws governing these categories are applicable to essentially all workplaces. Not all laws specifically target workplace safety but may have an indirect impact on enhancing workplace safety. Workplace laws that directly address and target workplace safety are mandatory to be implemented by all employers as they apply.

3.2 WORKPLACE SAFETY AND HEALTH LAWS

Among the plethora of laws that companies must implement and adhere to are the cadre of health and safety laws that places primary focus on the employees performing the various tasks that yield a service or a product in support of the goal of a company. These regulations are necessary and have proven to be beneficial in protecting workers from injuries and exposure to potential constituents or hazards that can produce illness while in the workplace.

One of the primary agencies that regulate health and safety in the workplace is the Occupational Safety and Health Administration (OSHA). In addition, the OSHA also provides safe work practices that are designed to be used in a wide variety of small and medium-sized business environments. The primary goal of occupational safety and health programs is to prevent workplace injuries, illnesses, and deaths from the workplace environment. These laws are also designed to prevent or minimize the suffering and financial hardship that these events can cause for workers, their families, and employers.

Employers will find that implementing these programs and practices will enable benefits such as:

- *Preventing* workplace injuries and illnesses
- *Improve* compliance with laws and regulations
- *Reduce* costs such as reductions in workers' compensation premiums
- *Enhance* worker engagement
- *Demonstrate* and enhance a company's social responsibility
- *Increase* productivity and enhance business operations[3]

The Occupational Safety and Health Administration covers health and safety laws that have been promulgated in the United States. Other agencies are actively enacting laws for other parts of the world such as the Health and Safety Executive which has responsibility for enforcing health and safety laws in the United Kingdom (UK).

3.2.1 Health and Safety Laws and Regulations

Safety and health laws and regulations are promulgated to ensure that workers can work in various work environments and return home in the manner in which they come to work each day. These regulations have significant importance to the company as well as the employees having responsibility to perform tasks and conduct business to assist the business in achieving their goals and objectives. These laws assist in helping to add sensibility to managing risks to protect the worker and the business and improve sustainability of the company.

Workplace health and safety laws are important for a variety of reasons. Some of these reasons include:

- Moral Correctness – It is morally right to ensure that workers can perform work activities in the work environment and return home in the same conditions that they reported to work on that day.
- Competitive Advantage – A company with good health and safety record can be viewed as having a competitive advantage in their respective business and industry. A good safety and health record builds trust in the company and adds to the perception of its reputation and brand. On the other hand, a poor health and safety performance can affect the profitability and sustainability of a business.
- Saves Money – Reducing downtime caused by illness and accidents means less disruption.
- Increase in Productivity and Efficiency – Protecting the workers can reduce absences; therefore, they are at work contributing to the goals of the company.
- Noncompliance Is Costly – In many countries health and safety regulations are legally binding. Therefore, employers must follow the requirements as documented in the regulations. Noncompliances can result in prosecution, fines, or imprisonment of company leaders.
- Many customers have a desire to conduct business with or purchase products from companies that operate ethically and ensure workers are protected during the performance of their work activities.

3.2.2 Environment Laws and Regulations

Environmental laws and regulations are not only designed to protect the environment and the public, but they also provide protection to workers in the work environment. Some of these regulations include:

- The Clean Air Act (CAA)
- The Clean Water Act (CWA)
- The Toxic Substance and Control Act (TSCA)
- Resource Conservation and Recovery Act (RCRA)
- Comprehensive Environmental Response Compensation and Liability Act (CERCLA)
- Federal Insecticide, Fungicide, and Rodenticide Act (FIFRA)

The list of possible questions to answer when determining what health and safety or environmental regulations (Tables 3.2 and 3.3) may apply to your company or business represents a short list of questions that can be used to spark and organize thoughts on compliance applicability and approach. This list is not comprehensive, and there are other considerations that should be made based on the business and the scope of the work being performed. Having knowledge of the regulations that

TABLE 3.2
Health and Safety Laws and Regulations Determination Aid

Questions

1. What is the scope of the organization in totality?
2. What is the scope of the regulation?
3. What is the applicable exclusion threshold?
4. What laws are applicable to all workplaces regardless of industry?
5. What is the mission of the organization?
6. What activities or products do the regulation apply to?
7. Are there any exclusions or exemptions that can be applied?
8. What potential hazards may be created during performing work activities?
9. What chemicals will be used in the workplace?
10. What are the requirements for reporting?
11. What are the recordkeeping requirements?
12. What chemicals will be created or developed as a part of performing work activities? If any
13. How many workers are employed by the company/organization?
14. What are the testing and monitoring requirements?
15. How many workers are employed by the operating departments?
16. Are there any enforcement provisions?

TABLE 3.3
Environmental Laws and Regulations Determination Aid

Possible Questions

1. What type of waste is expected to be generated as a by-product of conducting business?
2. How much waste is expected to be generated?
3. How will the waste be managed?
4. What inspection and reporting requirements need to be in place?
5. What procedures and policies need to be put into place to ensure and demonstrate compliance?
6. What type of air emissions are expected from business practices?
7. What type of environmental permits will be needed?
8. Who will be responsible for managing all permits?

are applicable to the business provides the basis for the development of a regulatory compliant program internal to the company to assist with maintaining compliance.

3.3 ENSURING AND MAINTAINING COMPLIANCE

Once the determination has been made on laws and regulations applicability the challenge is how to ensure that the company complies and remains in compliance. Maintaining compliance is not easy, and yes it does take dedicated resources depending on the regulations, the culture of the organization, and the work scope

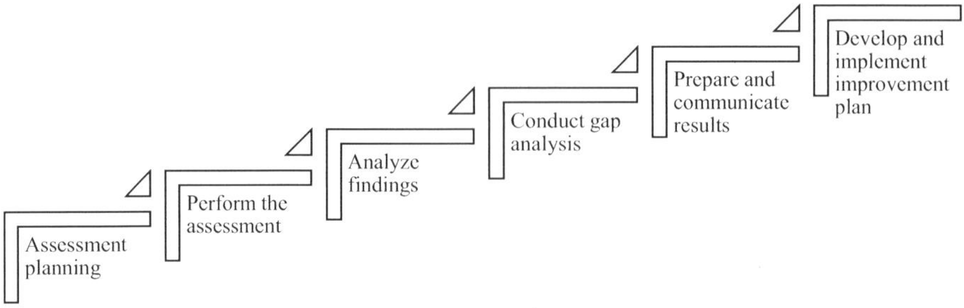

FIGURE 3.1 Compliance assessment process

performed. To assist in compliance maintenance most organizations have an individual or a group of people responsible for regulatory compliance. The size of the group depends on the number of regulations that are applicable to that business and the complexity of compliance. The compliance posture of the company should be evaluated at some frequency through activities such as procedure reviews, changes in regulations, and internal and external assessments. When assessing compliance, the steps listed in Figure 3.1 can be beneficial in charting the path to take to decision making and implementation based upon the information gained through implementation of the process. These steps are discussed in further detail in the Sections 3.3.1–3.3.6.

3.3.1 Assessment Planning

Planning is the first step in conducting a compliance assessment. It is paramount that before embarking on an assessment the scope of the assessment should be defined taking into consideration the following:

- The organization and its compliance goals
- Key Performance Indicators (KPI)
- Who will conduct the assessment
- Internal assessors
- External assessors
- The applicable laws and regulations

The assessment plan should address the key elements such as:

- The assessment criteria – define specific assessment criteria containing standards against which the assessment will be conducted. These criteria should be aligned with the compliance requirements
- Data collection – the methods to be used that include review of policies, procedures, and relevant documents as well as conducting interviews or surveys

- Data analysis – process used to identify patterns, trends, and potential areas of noncompliance for the data collected
- Conducting gap analysis – development of gap analysis and improvement plan
- Communicate findings – method used to communicate results and the population to which the communication is made. For written reports the report should include a brief description of the assessment process, areas of compliance and noncompliance, and any recommended actions included by the assessor

As a part of the preparation process ensure that you have buy-in from the leadership team and that the assessment is aligned with the company goals and KPIs.

3.3.2 Perform the Compliance Assessment

An effective compliance evaluation should assess compliance throughout the organization and identify a specific compliance area's strengths, weaknesses, and opportunities for improvement. Important components to take into consideration when conducting the compliance assessment include a review of policies and procedures that apply to each regulatory requirement being assessed; leadership oversight and their awareness of compliance posture of the organization; organization risk profile to include mitigation actions; employee training and engagement in organizational processes. Just as important are the team members that have been selected to perform the assessment and the size of the assessment team. The assessment team members should possess the required subject matter knowledge and regulatory knowledge to provide value to the assessment and comprehensively add to the evaluation of compliance and identification of any noncompliance within the organization. Additional skills one should seek in team members include the ability to communicate at all levels and listen to information as presented.

3.3.3 Analyze Findings

This step in the process involves analyzing the data to identify patterns, interpret findings or noncompliances, and gaps in compliance. It is during this step that any limitations of the data are also discussed. Examine data against what is needed to be fully compliant with the specific required regulations and standards that are applicable to the organization being deemed as compliant. The analysis can be used as a benchmark or baseline for when the organization completes the next compliance assessment.

3.3.4 Conduct Gap Analysis

A comprehensive gap analysis allows for the comparison of the current state and the desired state. Perform a gap analysis in a manner to identify the gaps between what is needed to be fully compliant and what the organization is currently performing.

During the analysis, the area of analysis is identified and analyzed along with the goals to be achieved; determine what the future state looks like; analyze the current state; compare the current state and the future state; document the gap; and describe the difference.

During the implementation of this step, it is paramount that the personnel involved in this process possess knowledge of the regulatory requirements and areas where the organization fell short in meeting its compliance obligations. An effective gap analysis will provide a map that can be used for a return to compliance posture and improve the effectiveness of programs.

3.3.5 Prepare Report and Communicate Results

Once the assessment has been completed and gaps identified, the findings should be reported to the organization and key stakeholders. Before distributing the report, ensure that the report is reviewed by others to verify accuracy and correctness. The report should clearly highlight the gaps and documentation of data provided. Solicit feedback from the leadership team and incorporate their feedback into a plan to improve the organization's compliance posture.

3.3.6 Develop and Implement Improvement Plan

Developing a plan to improve your compliance posture and to mitigate risk is the last step in assessing compliance. Rank each noncompliance and risk identified in terms of their business impact and risk level. Establish a centralized repository to document compliance tasks and evidence demonstration completion of each task. Each action item should have an assigned responsible party and completion date. Progress of the implementation plan should be assigned to an individual for monitoring completion of appropriate actions. The results from these evaluations should provide an indication of issues that may need to be addressed or changes needing in the current business practices. Once the assessment has been conducted, the improvement plan has been completed and issues resolved, it is important that the appropriate protocol be put into place to ensure compliance is maintained.

3.4 BENEFITS OF REGULATORY COMPLIANCE

Regulatory compliance is different from corporate compliance. Regulatory compliance represents compliance with laws and regulations that are mandated by the state, federal, or international government whereas corporate compliance can simply mean complying with policies and procedures that have been implemented internally by the company. Compliance with both is key in driving company accountability. Maintaining regulatory compliance helps an organization mitigate risks and actions that may include damaged reputation, loss of customers, and financial penalties.

3.4.1 Compliance Culture

Culture plays a significant role in organizations shaping the way work is performed, interactions with customers, and interactions between organizational members. Today it is more relevant than ever that a company develops and maintains a compliance culture because of the heightened regulatory enforcement environment. A compliance culture is not easy to define; however, there are some basic characteristics that must be engrained in a culture to provide the opportunity for a business to operate in the compliance realm. These elements are cyclical and must be present at all times to facilitate effectiveness. These elements can be placed in three buckets that consist of people, paper, and practices. Figure 3.2 shows these elements and their interactions toward a compliance culture.

These elements must be evaluated from time to time to verify and validate appropriateness and effectiveness. In doing so there are some key questions that can be asked in determining whether the culture where the culture is on a continuum from weak to strong. These questions include the following:

- What does a compliance culture look like for the respective business?
- Are there effective means in place to evaluate compliance?
- What involvement do workers have in facilitating a culture of compliance?
- Are the right policies in place to ensure compliance?

3.4.1.1 People

The actions of people form the foundation of a compliance culture. People are a significant factor in achieving work and building the foundation of a compliance culture. The people's aspects not only include workers but are highly connected and dependent upon the caliber and actions of the leadership team. All acts involving work in an organization have some connection to the workers in that organization.

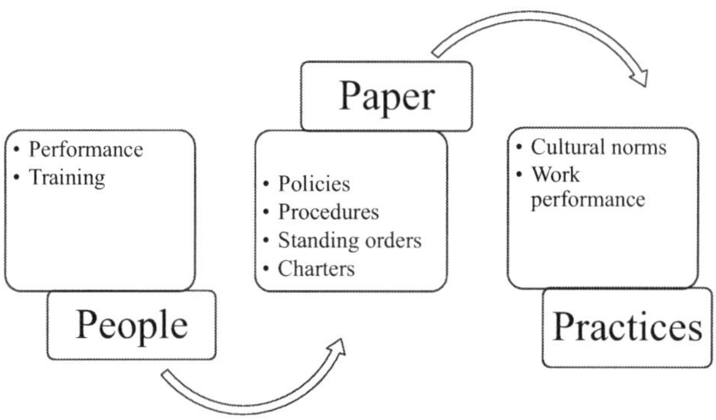

FIGURE 3.2 Compliance elements – 3 PPPs

3.4.1.2 Paper

The associated paper needed to document the expectations for members to follow can be minimal or massive depending on the business type and the number of regulations applicable to that business. The array of documents can include policies, procedures, documents associated with regulatory permits, committees, and team charters.

3.4.1.3 Practices

Work practices and cultural norms play an important role in a company's compliance culture. The culture of an organization dictates the work practices that the workers deem acceptable and whether they are compelled to engage in work practices that will ensure compliance to the applicable regulation requirements. In a culture where compliance is viewed as unimportant, favoring production and productivity, one can almost ensure that the complete compliance posture is not based on regulatory requirements and potentially based on what is viewed as easy and less costly. The fact is that easy and less costly in the near future usually results in complexity in compliance and more costly in the future. For example, if compliance is neglected until an issue is discovered during an audit, the cost associated with coming into compliance, settling potential fines and penalties, and loss of reputation of the company suddenly takes on a much greater cost than if the company were complying and at the onset and remained in compliance. Compliance will not become and remain a reality if the appropriate culture is not in place, supporting the appropriate practices expected of all who perform work in the environment expecting to be compliant. Culture and work practice are extremely important in an environment to facilitate safe work performance.

3.4.2 Failure of Determining Regulatory Compliance

Compliance cannot be achieved if an organization has no knowledge of the regulations that govern their workplace and industry. Ignorance of the law or rules is not a defense one can utilize when it comes to meeting compliance obligations.

At inception leaders should establish the company's compliance posture and be cognizant of the laws that govern their business. These laws often are many; therefore, it is necessary to have a dedicated group or person reporting to senior leadership responsible for facilitating compliance. In many companies this group is referred to as Regulatory Compliance having specific functions. The size of the structure of the group should complement the size of the organization and the number of regulatory requirements guiding the business. Determining and monitoring regulatory compliance is not an easy feat; however, it is necessary to protect the business as well as workers. A company's failure to determine the laws that are applicable to their business introduces additional risks that can severely impact the business. These impacts can range from tarnished reputation and difficulty in attracting and retaining workers to increased regulatory oversight and fines.

Although the cost of compliance continues to rise, compliance can be more costly if a company is found to be out of compliance and faces the associated fines and penalties. The rising cost of compliance presents challenges for small businesses to come into and remain in compliance. However, business size is not a defensible reason for not determining regulatory applicability. Determining regulatory applicability and maintaining compliance with those laws and regulations is a requirement for businesses to achieve and maintain compliance with regulations.

3.5 WHISTLEBLOWERS

Organizations that operate in a state of noncompliance have the potential to have employees who will contact the appropriate regulatory agencies and provide information on the noncompliant business practices. These employees are referred to as whistleblowers. It is acceptable for employees to report noncompliances to regulatory agencies when they have knowledge of their occurrences. When an employee becomes a whistleblower, it is generally because they believe that they have no other recourse to impact change within the organization. Whistleblowing has become a common occurrence in today's corporate environment. Whistleblowing is the act of reporting noncompliance, unethical, or illegal activities that occur within an organization by an employer or stakeholder.

REFERENCES

1. OSHA 30th Anniversary, https://www.osha.gov/aboutosha/30-years
2. Laws and Worker Protection, USA Gov, https://www.usa.gov/labor-laws
3. Occupational Safety And Health Administration, Recommended Practices for Safety and Health Programs, A safe Workplace is Sound Business, https://www.osha.gov/safety-management

Part II

Chapters 4–6 focus on the human aspects that must be considered and channeled in developing and implementing safety and health programs to provide the culture that underpins a safe and inclusive workplace.

DOI: 10.1201/9781032624389-5

4 Human Performance and Behavior Modification

4.1 THE IMPACT OF BEHAVIOR ON WORKPLACE SAFETY

Employees are an important part of an organization being able to achieve its goals and deliver for their customers and stakeholders. With a diverse population of employees comes a wide array of thoughts and behaviors that must be integrated, modified, or accepted. Many of these behaviors can be counterproductive to the goals of the organization and the culture they are seeking to develop and maintain. It is also recognized that as humans it is expected that mistakes will be made. These mistakes depending on the severity and their impact can be costly to the workers, the business, and the neighboring communities. Therefore, it is incumbent that an understanding be gained as to how, why, and when mistakes are made to minimize them. Having the answers to these questions, leaders can focus on error modification through designing processes, developing programs and procedures, instituting, or modifying work practices that can minimize the potential for errors.

The behavior of workers has a direct impact on the occurrence of workplace errors. Human behavior refers to the way an individual presents themselves through their actions and interactions. These actions and interactions include what is communicated verbally as well as nonverbal.[1] Some core assumptions of human performance include the premise that:

- All humans are fallible – everyone given the right set of circumstances and environment will make mistakes
- The behavior of individuals in the workplace is influenced by the work environment. The work environment encompasses the culture, the actions and interactions of management, and the policies and procedures that guide work
- Error-likely situations and conditions are predictable and preventable through the recognition of hazards that contribute to the creation of errors. Identification and removal of hazards is the key to preventing errors
- Encouragement and reinforcement of desired behavior impact the level of performance of workers. Organizational safety and health are directly related to the behaviors of humans
- Understanding the causes of mistakes provides knowledge needed to avoid the resultant error. Evaluating and anticipating the impacts of past errors and their impact on future activity will assist in mistake avoidance

DOI: 10.1201/9781032624389-6

Keeping the aforementioned assumptions and many other assumptions that are not mentioned can impact human performance, in mind will provide insight into and aid in the prevention of mistakes that can lead to errors while maintaining a focus on the actions and performance of workers.

4.1.1 People Make Errors

Studies have shown that managing human error can yield a positive return on investment for companies. This return on investment can come in the form of savings incurred from not having workers away from work, penalties that may be levied as a result of failure to comply with laws and regulations, or penalties that may impact the business in attracting and maintaining skilled workers. Recognizing that humans possess a wide range of capabilities as well as shortfalls is an important attribute in identifying potentially fallible behaviors modifying behaviors and reducing errors. Modifying the work conditions to accommodate the existence of error traps when performing a task, or at least minimizing the chance for error, can prevent a significant injury-causing event. Human behaviors observed within an organization can be a direct reflection on the culture of that organization, the leadership team, and the policies, practices, and procedures that guide work.

All behaviors, whether safe or unsafe or described as bad or good behavior, are reinforced by experiences, modeling what has been witnessed, or direct is a consequence of what was experienced. Many workplace behaviors are underpinned by the consequences that the employee experiences when the behavior occurs. People tend to evaluate behaviors they see and determine what is acceptable and then model it. For example, if an organization allows taking shortcuts, rude and distracting behaviors, and procedure noncompliances, then the employees will view those behaviors as acceptable and emulate them. As such, these behaviors become engrained in the way work is conducted, often leading to errors.

Depending on the reports and studies reviewed, error rates for humans can vary from between one and fifteen errors per hour. In fact, one study documented that people, when performing any activity, generally commit on average three to four errors per hour. Regardless of which study one chooses to agree with, the fact of the matter is that a focus on error prevention is critical for companies to achieve their goals and sustainability of the business. Not all errors have bad outcomes, some errors have led to great discoveries, for example, "Post It" notes were developed during the plight to invent a super strong adhesive. Although the error that led to the discovery of Post-It Notes represents a good story, most errors result in unwanted outcomes such as an injured worker or system upset. These unwanted outcomes can be catastrophic and significantly set a company back in many ways.

Errors will mostly fall into one of two categories, active and latent. An active error is an observable error resulting in immediate adverse outcomes. Front-line workers are usually the ones that commit these types of errors because they are in the environment where these errors occur and have hands-on responsibility for accomplishing work. An example of an active error would be manipulating the incorrect valve on a system diverting products in an unwanted location or configuration or turning

TABLE 4.1
Attributes of Error Types

	Active Error	Latent Error
Who generally makes these types of error?	Front-line hands-on Workers	Administrative workers, procedure writers, managers, support staff
What caused the error?	Equipment	Documents (procedures, policies, work instructions), values
When is it evident that an error occurred?	Immediately	Later, delayed
Is the error visible at occurrence?	Yes	No

off a safety switch that shuts down a piece of an equipment if it is unsafe to operate. An active error will generate a noticeable undesirable consequence. Latent errors appear as weaknesses in processes, and procedures, as inefficiencies, and unwelcome changes in practices. These mistakes are typically documentation-based, such as incomplete procedures, policies, drawings, and design documents. Error occurs as a result of workers utilizing this incomplete documentation. Active errors appear immediately, and latent errors are not readily apparent and may take days, weeks, or years to be discovered. Table 4.1 provides some examples of the two types of errors.

4.1.2 Human Characteristics

The characteristics each person possesses become a part of the work environment once they join a company or a work team. These characteristics combined with the characteristics of other workers have a guiding impact on the formation of the work culture, the interaction between workers and the management team, and the performance of work. Common traits possessed by humans such as personality, and restraints may bias workers to err under adverse conditions. These traits include behaviors, short-term memory, stress, complacency, inaccurate view of risk, and reduced recall of information. Examples of how these traits may impact a worker and the workplace are depicted in Table 4.2.

A significant trait that has the ability to derail a workplace is stress because it represents a significant distraction for workers as it can contribute to poor decision making, loss of attention and focus, and impairment in the ability to recall key facts, thus increasing the potential of making an error. Individuals handle stress differently and have different tolerance levels in the amount of stress that they can shoulder without it becoming impactive to work and home life. In addition, workers must deal with stress off and on the job and at times may feel unbalanced or overwhelmed. These feelings may manifest themselves in the behavior of an individual, modifying the traits that would commonly be expected from that individual.

Human characteristics can lead to the ignoring of changing conditions or process variances because of conformance to routine protocols. As workers become more

TABLE 4.2
Impact of Some Common Traits

Trait	Potential Workplace Impact
Behaviors	Disruptive behaviors in the workplace can negatively impact the performance of the disrupter as well as other employees. These behaviors can also lead to low morale and lack of trust
Short-term memory loss	Significantly reduce the effectiveness of training and an ability to perform tasks requiring sequencing of work steps
Stress	Reduce focus and attention to detail resulting from stress can cause a worker to perform work in an unsafe manner or produce poor quality work products
Complacency	A reduction in an employee's questioning attitude that can result from complacency can lead to performing a routine task without noticing a minor change in process or steps that can result in injury or failure in a system
Inaccurate view of risk	A lack of knowledge or respect for risk can lead to decision making that leads to system failure or injury to workers
Reduce recall of information	Significantly reduce the effectiveness of training and an ability to perform tasks requiring sequencing of work steps

experienced and continue to perform the same task day-in and day-out achieving the same results, they can become complacent or myopic and overlook changes that may occur within a system or process that has created a changing condition resulting in a hazard. They may also begin assuming that current-day conditions will always be the same, not having adequate situational awareness and invoking their questioning attitude each time a task is performed. Preconceived notions can cause employees to become myopic and overlook information/data that is critical to decision making that can eliminate an error. Evaluation of hazards can depend on an individual's perceptions and knowledge of the task or system and their degree of risk taking. Worker knowledge and level of risk taking coupled together or invoked separately can facilitate errors and limit their understanding of the associated risk and potential impacts.

A study published in *Frontiers in Psychology*, on March 28, 2023, suggests that human traits and employment status are related. This study further stated that there is a significant relationship between personality and job performance.[2] Understanding the connection between these characteristics that we possess as humans that form one's personality and guide behaviors may provide insight into how we perform in the workplace. Personality traits are relevant to harmful and problematic workplace behaviors since personality traits relate to and may shape the attitudes and behaviors of individuals, as well as impact their functioning as part of a team or organization as a whole.[3] The understanding and connection between an individual trait and behavior can be useful in developing, implementing, and maintaining a safe work environment.

4.2 WORK CULTURE BEHAVIORAL IMPACTS

Developing the appropriate culture for a work environment comes with many challenges. The many challenges are represented in the wide array of books and articles written on culture in the workplace. A significant part of developing and maintaining culture resides with the people that are operating within that culture. As such, the behavior of workers and management are important as they bring with them differing behaviors that must be assimilated into their workplace environment. The differing behaviors make it difficult to facilitate one culture, avoiding the presence of a variety of subcultures within the work environment. Some of these behaviors are positive and some are not so positive; some are beneficial, and some are destructive.

It is recognized by leaders that culture has far-reaching impacts on organizations ranging from the way work is being performed, relationships with customers, procedure implementation, and integration of a diverse workgroup with diverse backgrounds. To gauge employee behavior and the culture of the workplace, surveys are distributed by many employers, some on an annual basis to gauge behavioral impacts and culture status. The data/information gleaned from these surveys are used to modify programs, policies, procedures, and training as needed to continue to address issues and potential issues and improve behaviors and the culture of the work environment. In addition, many employers implement training such as behavior-based safety and culture as a requirement for all employees including management.

4.2.1 Leadership and Managers' Behavioral Impacts

Leaders and managers play a significant role in shaping an organization and the work environment. They have the ability to set the stage and path that will dictate the way the organization responds to all aspects of the business, including performance and the behavior exhibited by members of that organization while performing their work activities. Leaders are ultimately responsible for charting the path of an organization in addition to modeling the behavior that is expected and desired of others. Today, some workplaces are viewed as dysfunctional and unproductive by workers because of the behaviors of their management team. Some contributions of leadership/management that can positively impact behaviors include:

- Following all procedures, policies, and prescribed work practices. Leaders and managers that set the example by following policies and procedures have the ability to compel and influence organization members to follow their lead
- Serving as role model through display of respectful behaviors. Respectful behaviors exhibited by the leadership team have the ability to influence others to treat each other with respect to include respecting the differing opinions and ideas of others
- Exhibit trustworthy behaviors in action, practices, and the policies and procedures implemented on behalf of the organization. Leaders that are trusted

can exert influence and compel workers to support organizations' goals and objectives
- Demonstrate concern for employees through facilitating a safe work environment and demonstrate valuing their opinions and input

In addition to the contributions listed above, managers and leaders can help shape culture through the policies and procedures they put in place, their actions, reactions, and interactions with organization members. Some actions that managers can engage in to help shape the culture that can guide behaviors include demonstrating and promoting team concepts, consistently seeking feedback from others and acting on the knowledge gained, actively listening to workers, modeling the expected behavior, and demonstrating concern for employee health and safety.

4.3 BEHAVIOR MODIFICATION TECHNIQUES

To further enhance workplace safety, behavior modification should be considered and used where practicable. Before considering modification of behaviors there must be a clear understanding of the process, the available tools and techniques that can be used to render the process effective, and the potential impact of the various workplace laws to ensure compliance and effectiveness.

Behavior modification commonly refers to the techniques used to facilitate the decrease or increase of a specific type of behavior. The techniques associated with behavior modification have been used to some degree successfully in areas such as the healthcare industry as well as the educational arena. To date, this technique has not been used in the workplace as a means of rendering workers more effective in improving workplace safety and health. The concept of behavior modification relies on the concept of conditioning a form of learning.

There are two major types of conditioning: classical conditioning and operant conditioning. Classical conditioning relies on a particular stimulus or signal. An example of this would be if a family member came to the kitchen every time, they smelled a specific food being prepared. Another example of a classical conditioning is when a worker develops a heightened anxiety response to a specific worksite location after experiencing a near-miss or accident there, causing them to feel nervous even when no immediate danger is present, simply by being in that area again; the worksite location becomes the conditioned stimulus associated with the negative experience (unconditioned stimulus) of the near-miss, triggering a conditioned response of anxiety. The second type is known as operant conditioning, which involves using a system of rewards and punishments. An example would be rewarding a dog during training with a special treat after they obey a command.[4] Another example of operant conditioning is using a safety app to reward workers for consistently following safety policies and procedures

The purpose of considering behavior modification is to seek means of changing behavior that are not considered positive. There are several different methods that can be used to accomplish behavioral changes. These methods are included below:

- Positive Reinforcement – is the process of adding a positive stimulus to the behavior. An example of this is when teachers reward students for getting a good grade with a prize. Another example of positive reinforcement would be a supervisor verbally praising an employee for consistently wearing their safety goggles on the job site, thus encouraging them to continue this safe behavior by acknowledging their actions positively.
- Negative Reinforcement – involves pairing of behavior to the removal of a negative stimulus. For example, a child that throws a tantrum because they don't want to eat vegetables, therefore, the vegetables are taken away. Another example of negative reinforcement is when an employee consistently wears safety goggles to avoid receiving a reprimand or disciplinary action from their supervisor for not following safety policy; they are performing the desired behavior (wearing goggles) to avoid the negative consequence of being reprimanded.
- Punishment – is designed to weaken behaviors by applying an unpleasant stimulus to the behavior. Instating detention for bad behavior is an example of punishment.
- Flooding – involves exposing people to fear involving objects or situations intensely and rapidly. For example, forcing someone who fears an object to hold that object for a period of time.
- Systematic Desensitization – is also used to treat phobias and involves teaching an individual to remain calm while focusing on their fears. For example, a road traffic worker with a fear of bridges might start by looking at a photo of bridges, then thinking about standing on a bridge and walking over a real bridge, perhaps starting with a small bridge.
- Aversion Therapy – involves showing love toward a person while focusing on positive thoughts about the person. Noticing significant good aspects of the person possesses as opposed to finding fault.
- Extinction – involves forcing a person to get rid of unwanted behaviors. For example, a child that is put on a timeout because of unwanted display of behavior may eventually stop that behavior because they do not want to be placed in time out.[4] Another example of extinction could be a workplace where employees frequently disregard a safety policy by not wearing required safety glasses, and management decides to stop verbally reprimanding employees who are not wearing their safety glasses, but instead focusing on positively reinforcing employees who are consistently wearing them, effectively stopping the unwanted behavior of not wearing safety glasses by removing the attention previously given to it.

Table 4.3 presents potential ways some behavior modification techniques may be considered to be used in the workplace. Before embarking upon using these techniques ensure that the legal and medical aspects are covered, and the appropriate policies and procedures are in place.

TABLE 4.3
Behavior Modification Techniques Potential Application in the Workplace

Behavior Modification Method	Potential Application in the Workplace
Positive reinforcement	Providing recognition for accomplishing activities safely
Negative reinforcement	May not be applicable or embraced in the workplace as it can conflict with workplace laws designed to protect the rights of employees
Punishment	Revoking a decision that once was made by workers requiring a higher level of management to provide direct supervisor monitoring. This action may typically take place after a safety occurrence to avoid reoccurrence. Although not viewed as punishment by management may be viewed as punishment by workers
Flooding	May not be applicable or embraced in the workplace as it can conflict with workplace laws designed to protect the rights of employees
Systemic desensitization	Can be applied in the workplace for workers that have to work at height or climb ladders or scaffolds that are uncomfortable with the idea. This technique must be applied carefully and ensure adherence to workplace employee laws.
Aversion therapy	Focusing on the good behaviors of workers
Extinction	Subjecting the worker to a penalty when a persistent unwanted behavior is witnessed.

Behavior modification, also referred to as behavior management and behavior analysis, focuses on changing an individual's environment in a manner that increases the likelihood that specific behaviors will occur. Behavior modification focuses on reinforcing behaviors so that they occur more often. An important factor that often gets overlooked with behavior modification is change from the perspective of decreasing undesired behavior and increasing desired behaviors. Behavioral strategies are not supposed to remain the same. When you find one type of approach that works, you should keep it in place for a short period of time and then adjust the program. An effective program will cease to remain effective if left unchanged. The fact is that behavior modification programs are supposed to change.

An important complex part of behavior modification is choosing what reinforcement to use. There are many types of reinforcement, and not all of them will work for everyone. Some of the most common types of reinforcement include food, paying attention, avoidance of something a person does not want to do, engaging in fun things, and money. Everyone responds differently to reinforcement; thus, each behavior modification plan should use different reinforcers. If a behavior management plan is not working, a possible reason is that the type of reinforcement that is being used is not effective. The following can be helpful when implementing a behavior modification strategy:

- Behavioral plans are expected to change and not remain the same. Once it is known that a plan is working, some modification may be needed so that the reinforcement occurs on a different schedule and in a complementary manner. This action is instrumental in keeping a plan effective over time.
- Because people are different, not every type of reinforcer will work for all individuals. The first step in developing a behavior management strategy is to decide what reinforcer is most likely to be effective.
- People tend to respond better to approaches that they expect will bring about certain outcomes; therefore, consistency is very important for effective behavior modification plans.
- Various people with different personalities will often have different impacts on the effectiveness of a behavior plan. If a particular plan seems to work well for a particular individual, then try to identify what that person is doing so that the approach or similar approach can be used for others.[2]

There are several types of reinforcement that can impact behavior. Some of these reinforcements may not be appropriate for the work environment. When a behavior plan is initially set up, continuous reinforcement can be used to establish and reinforce the behavior. Continuous reinforcement can change to intermittent reinforcement which is referred to as thinning after the behavior has been established. There are four types of intermittent reinforcement available for use. They are listed below in Table 4.4.

Behavior modification techniques may have positive implications for the work environment to improve worker health and safety performance, compliance with

TABLE 4.4
Behavioral Reinforcement

Intermittent Reinforcement Type	Reinforcement	Examples
Fixed interval	The person is reinforced by a set number of responses	Rewarding a person at the end of each day
Variable Interval	The person is reinforced by a variable number of responses	Rewarding a person sometimes at the end of the day, sometimes at the end of the week, sometimes every few days
Fixed ratio	The person is reinforced after a certain number of responses	Rewarding a person after completing the desired behavior four times
Variable ratio	The person is reinforced after a variable number of responses. Variable ratio intermittent reinforcement is the most effective schedule to reinforce a behavior	Rewarding a person after completing the desired behavior after three times, then after six times, then after two times. Gambling is a real-world example of a variable ratio of reinforcement

Source[3]

procedures, and work performance. This technique may serve as a supplement notwithstanding the potential challenges that may be encountered with the varying methods of a behavior-based safety program that are in place at many workplaces.

4.4 HUMAN PERFORMANCE IMPROVEMENT PROGRAM

Human performance is the behavior (B) plus results (R) equals performance (P), B + R = P. Behavior is what people do and say, its actions that can be observed and seen by others. All behavior, whether good or bad, is reinforced by experiences and direct consequences.

In the work environment, there is a desire to achieve accelerated productivity and produce a service or product that is pleasing to customers without injuries to workers during work activities. It is well documented that people make errors that can cause significant events that can lead to injuries and death. Successfully managing human errors can yield benefits such as fewer injuries and unwanted events, reduced downtime, greater productivity, and increased efficiency.[1]

Attributes of an effective Human Performance Improvement Program (HPIP) should include the elements that are listed in Figure 4.1. These attributes are the minimum requirement that should be included in the HPIP to achieve some level of effectiveness and success in charting behaviors while reducing systemic errors.

4.4.1 Understanding Error Precursors

There must be a keen understanding of error precursors in order to prevent errors from occurring. The first question at hand is what is an error precursor? An error precursor is an adverse prior circumstance that can increase the probability of

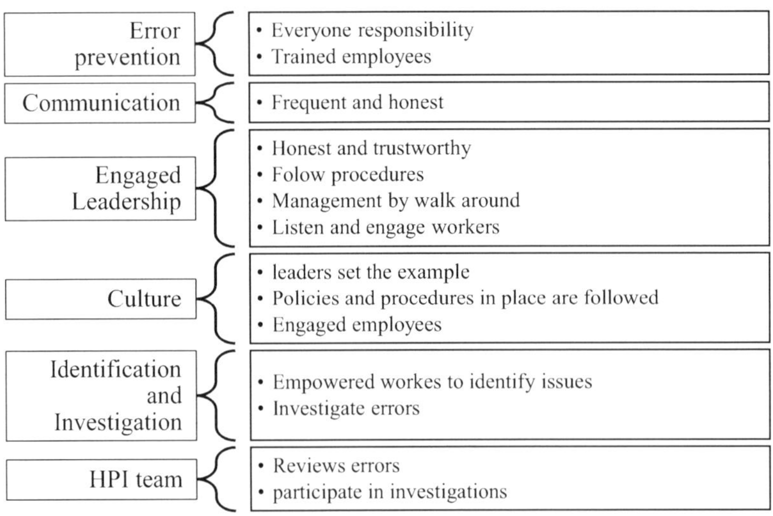

FIGURE 4.1 HPIP elements

TABLE 4.5
Error Precursors Examples

Task Requirements	(Work Setting) Potential Pitfall
Excessive workload	Distractions, disruptions
Rushing – time pressure	Unclear visual instructions, or exhibits
Multi-tasking (working simultaneously on more than one task)	Equipment that is not functioning to its fullest capacity, finding time and alternate ways to complete the task
Clarification of request or instructions	Lack of understanding of equipment status
Unclear purposes, functions, accountabilities	No additional indication or understanding of status
Unclear criteria	Individual struggles or comprehension
Employee Competencies	**Human Characteristics**
First time to complete activity, inexperienced	Anxiety, stress, unsure of abilities
Limited in-depth understanding of task	Routine protocols
New method has not been utilized before	Theoretical, preconceived notions
Poor communications	Aggressive, complacency
Limited ability to resolve issues/problems	Flawed view of risk
Lackadaisical attitude toward important actions	Preferences, philosophical shortcuts
Low energy, tiredness, illness	Reduced ability to think clearly or recall information

making a mistake. These circumstances typically occur when the requirements of a task exceed the capability of the person performing the task or when the work settings exacerbate the boundaries of the employee's abilities to respond appropriately and timely. These predictable mistakes or errors can be grouped predominantly into four categories, namely, task requirements, work settings, employee competencies, and human characteristics. Table 4.5 provides some examples of each type of error.

REFERENCES

1. Allen, P.M., Alston, F.E., and Dekerchove, E.M. *Peak Performance How to Achieve and Sustain Excellence in Operations Management* (1st ed.), CRC Press, Taylor & Francis Group, 2019.
2. Kang, Weixi, Guzman, Kreisha, Malvaso, Antonio. Big five personality traits in the workplace: Investigating personality differences between employees, supervisors, managers, and entrepreneurs. *Frontiers in Psychology*, 14, 2023. https://doi.org/10.3389/fpsyg.2023.976022
3. Preston, O.C., Anestis, J.C., Watts, A.L. *et al.* Psychopathic personality traits in the workplace: Implications for interpersonally- and organizationally-directed counterproductive and citizenship behaviors. *Journal of Psychopathology and Behavioral Assessment*, 44, 591–607, 2022, https://doi-org.oregonstate.idm.oclc.org/10.1007/s10862-021-09918-8

4. Vijayalakshmi, N. Behaviour modification techniques: An awareness study. *Shanlax International Journal of Education*, 7(2), 20–24, 2019, https://doi.org/10.34293/education.v7i2.333
5. Psychology Today, Understanding What Makes Behavior Modification Works, October 2, 2018, https://www.psychologytoday.com/us/blog/comparatively-speaking/201810/understanding-what-makes-behavior-modification-work
6. Scott, H.K., Jain, A., and Cogburn, M. Behavior Modification. 2023 Jul 10. In: *StatPear*ls [Internet]. StatPearls Publishing; 2023 Jan–. PMID: 29083709

5 Culture Evaluations and Modifications

5.1 ORGANIZATIONAL CULTURE

Culture within organizations has been studied for decades, and the result of extensive studies has concluded that culture is an important aspect of an organization that guides the function of an organization. The quantity of these studies suggests that culture can be created to exhibit the creator's values, priorities, and vision. In addition, these studies express that culture contributes to attributes such as employee commitment, productivity, and corporate profitability. Important components of a great culture include a vision and a mission that are supported by members of the organization and values that are clear, sound, and communicated. Documentation is in place that clearly identifies the guidelines for expected behaviors; sound practices that demonstrate management's value for their employees and customers and a leadership team that "walk the talk"; and employees that are engaged.

The culture of an organization serves as the foundation of the organization and enhances the ability for members to build relationships and trust among members. Human performance is significantly dependent on culture since culture has an influence on members in terms of releasing creativity and productivity. The culture of an organization drives the behaviors of the employees as they interact with each other and perform their job responsibilities. According to Alston 2014,[1] based upon an extensive literature review, there are four primary cultural attributes that are important in defining and sustaining the culture of an organization that include practices, values, beliefs, and behavior. Additional details and definitions of these attributes are shown in Table 5.1.

5.2 BUILDING AND MAINTAINING A HIGH-PERFORMANCE ORGANIZATION

Building and maintaining a high-performance organization is the desire of leadership because high-performance boosts quality and revenues. A high-performance organization is defined as an organization that can perform at a very high level in the market in which they compete fulfilling their vision and mission. In order to build a high-performance organization, it is necessary for the culture to be structured in such a manner to provide the fuel for success. Recognizing that culture building takes time; therefore, building a high-performing organization will take time. Nonetheless management must be motivated and invest the time and resources needed to build and sustain a culture capable of shepherding the organization into a

DOI: 10.1201/9781032624389-7

TABLE 5.1
Cultural Attributes

Attribute	Description
Belief	Substantial influence can be placed on an organization by the beliefs that are present within an organization because the engrained beliefs represent important variables that guide behaviors. This influence can have an impact on how the organization operates and the interactions between employees. If the leadership team demonstrates their belief and value in a compliant culture, then the expectation is that their subordinates will adopt that same beliefs and values. Because beliefs are ideas and ideals that people hold as being factual, compliance can be impacted when accurate beliefs are acted upon.
Behavior	Human behavior can be destructive or constructive and can reinforce the need for compliance or discourage workers from performing work in a compliant manner. There are three aspects of human behavior that can significantly impact compliance and culture. These aspects are people tend to follow leaders that they have confidence in; people seek unity and tend to follow others to avoid peer pressure; people react and respond to pressure. Considering these aspects as they are operating in a compliance culture leader must demonstrate their support and expectations for employees to perform work according to procedures, policies, and regulatory requirements.
Practice	Practices are the unwritten activities and ways of conducting business based on past rules or ways of thinking. These practices are ingrained by workers that have been with the organization for a long period of time. Ingrained ways of thinking and conducting business are extremely difficult to change in many cases, and these practices are handed down to new workers when they enter the organization. Although newly assigned workers are aware of the desired way of conducting business, they may succumb to the practices that are performed due to peer pressure or the desire to fit in and not be singled out as someone who has the desire to be different.
Values	Values refer to the standards that are routinely deemed as acceptable or unacceptable in an organization. Cultural values directly impact human interactions and behaviors that will have a direct impact on the propensity to place compliance needs on the same playing field as getting work accomplished. Organizational members must place value in ensuring that the organization develops and follows procedures and practices that will ensure compliance with all requirements. Shared values consist of the goals that are shared by people in an organization, that shape and guide behaviors. The behavior exhibited by people that are directed toward following policies and procedures has an impact on compliance.

higher realm. Section 5.3 discusses methods that can be used in building and maintaining the appropriate culture thus influencing the performance of organizations.

5.3 BUILDING A HIGH-PERFORMANCE CULTURE

Building a high-performance culture takes leadership involvement while exhibiting the appropriate behaviors and a strategy to build, maintain, and evaluate culture periodically.

High-performing organizations invest in and depend greatly on their culture. A high-performance culture consists of attributes such as:

- Continuous improvement mindset – Leaders possessing a continuous improvement mindset are always seeking ways to improve efficiency and quality.
- Leaders are able to inspire and motivate – Leaders have significant influence on the organization as well as employee performance. The actions and behavior of leaders set the stage for acceptance of the company values, goals, and mission and guide the behaviors exhibited by other members of the organization.
- Employees feel empowered and are engaged in all aspects of the business. Employees that feel empowered are engaged in problem solving and exerting their creativity to assist in the development and production of new processes, procedures, or products.
- Concepts of high-performance teams are actively used. High-performance teams are important in new ideas development that can lead to novel products and procedures for accomplishing work and improvement of product quality.
- Trust between management and employees – Organizational life is enhanced when trust permeates throughout and workers are willing to go the extra mile to ensure that the company goals and objectives are met.
- Lean thinking leaders – These leaders are actively engaged in using lean tools to improve quality and efficiency.
- Company performance is measured accurately and shared and is in alignment with the company strategy.

In a high-performing culture members of the organization are working together cohesively to ensure that the goals of the organization become a reality. Some benefits of a high-performing culture are listed in Figure 5.1. Each of these benefits is discussed in more detail in sections 5.3.1–5.3.4.

High-performing cultures yield organizations that are able to retain top talent, able to continuously improve performance, facilitate innovation and creativity, and able to retain satisfied customers. These benefits are explored further in sections 5.3.1–5.3.4. These benefits provide an incentive for managers to work toward and maintain cultures that support high performance.

5.3.1 Top Talent Retention

Employee retention is an important tool that high-performing culture excels in. It is not easy to attract and retain highly skilled employees; however, these organizations put in place tools and processes that will assist in this endeavor. Organizations cannot achieve a culture to support high performance without having on board the skilled staff needed to facilitate their vision and mission. Figure 5.2 shows key components

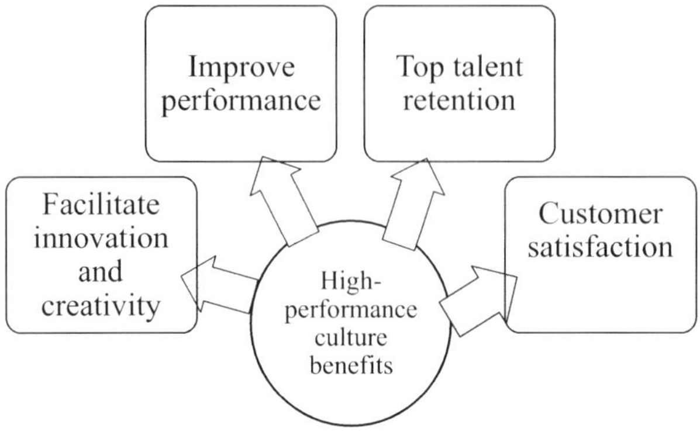

FIGURE 5.1 High-performance culture benefits

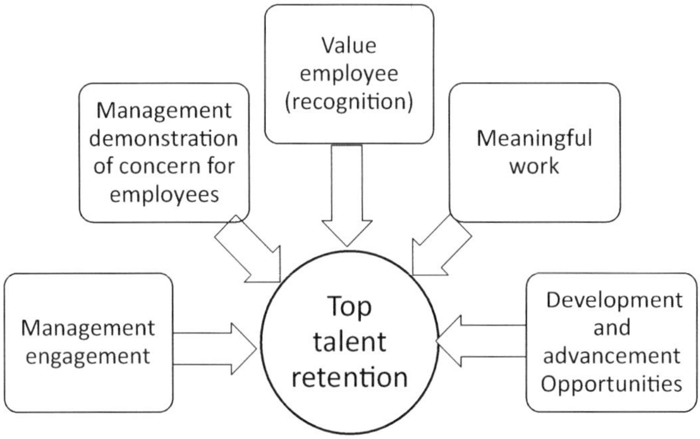

FIGURE 5.2 High-performance talent contribution

of top talent retention. These components are discussed in additional detail in sections 5.3.1.1 through 5.3.1.5.

5.3.1.1 Management Engagement

Management and leadership engagement is necessary for all organizational activities and processes to reach the level of success needed to render an organization effective. When it comes to attracting and retaining top talent, management must have a strategy for success and be at the forefront of the process. Management plays a critical role in worker retention and engagement which goes hand in hand in formulating the perception that workers have of their organization and to some extent the leadership team. To facilitate their roles effectively management should seek to build relationships with the subordinates, provide open and honest feedback frequently, facilitate trust, develop and encourage a positive work culture where employees feel free to communicate their thoughts and ideas, and support employees in their professional development efforts.

5.3.1.2 Management Demonstration of Concern for Employees

Concern for employees deals with having the feeling of caring, empathy, tolerance, and safety. It is important for management to listen to and gain an understanding of the perspective of others and provide open honest feedback during each interaction. During interactions with workers there is an expectation that managers are respectful of people and their ideas even when there is a fundamental differing of opinions. This display of concern serves as encouragement to workers that will propel them to engage in relationship building and formulate the bond between management and workers that is necessary to move a business forward and enhance productivity. In addition, workers that have the perspective that management is concerned about their well-being are more likely to support the organization's goals and objectives presented by management.

5.3.1.3 Employee Value

Employees feel valued when their involvement is sought out by management and their peers to exert their knowledge and opinion when solving problems. Therefore, many employers institute employee recognition programs as an integral part of their business strategy to serve as encouragement for employees to become actively engaged in problem solving. Employees also feel valued when they are recognized by management for their contribution to the success of the organization, a project, or a task. Recognition can be presented in various forms such as verbal, written, or financial. Some of the benefits of employee recognition and demonstrating to employees that they are valued include increased productivity, boosted morale, improved team cohesion, behavior modification, and acceptance of the values of the organization.

5.3.1.4 Meaningful Work

Employees that view their role in an organization as being meaningful are more likely to remain with the organization as long as they feel that the work they perform is meaningful and enjoyable. Recognizing that meaningful work may be defined differently by various employees, management must have a relationship and a connection with their direct reports so that they can understand their definition of meaningful work. Once knowledge of what constitutes meaningful work is known, it is beneficial for management to work with employees and structure work activities to meet employees' perception of their work responsibilities. An employee's perception of their role in the organization can directly impact their contribution to the success of the organization and willingness to remain with the organization long term.

5.3.1.5 Development and Advancement Opportunities

Employees when entering the workplace generally have a plan for career movement that includes expanding their knowledge and potentially upward mobility. They are seeking opportunities to develop, improve, and sharpen existing skills and be trained on mastering new skills. Providing opportunities for workers can build a talent pool that supports the organization's mission for the present and into the future. In addition, development opportunities can create engaged employees and improved performance. It is up to management to develop and institute programs and policies that will afford workers the opportunity to develop and advance in their careers and expand their knowledge. Career development and advancement opportunities may be the bridge needed to retain talented workers.

5.3.2 Improved Performance

Improved performance and profitability are seen in high-performing cultures primarily due to the engagement of high-performance teams and employees that are actively engaged in ensuring that the goal and mission of the organization is accomplished. Improved performance becomes a reality as shown in Figure 5.3, when the elements listed are engrained in the organization structure and culture. Effective leaders are engaged in all aspects of the organization, and trust is evident between management and employees.

The attributes depicted in Figure 5.3 that are important to achieving improved performance are discussed in more detail in sections 5.3.2.1 through 5.3.2.3.

5.3.2.1 Employee Engagement

Employees that are engaged in work activities and feel empowered tend to perform at their highest level or full potential. Engaged employees are also motivated to perform work while sharing their skills and knowledge with others. High employee engagement increases performance because the employees are more willing to go the extra mile, work to overcome obstacles, and solve problems. Engaged workers in a high-performing environment care about what they do at work and not just showing up for a paycheck. Employees that engage in business activities tend to be willing to look out for their safety and the safety of their coworkers while performing work.

5.3.2.2 Trust Permeates Throughout

People tend to follow those they trust even when they are not in total agreement with their philosophy or direction. Trust building and maintenance are the responsibility of the leadership team. Some activities/behaviors that leaders can engage in to build and sustain trust include:

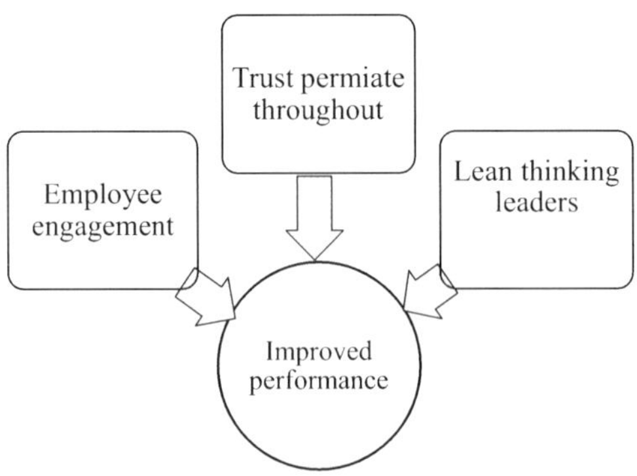

FIGURE 5.3 Improve performance impact

- Freely admit mistakes and use them as teachable moments
- Walk the talk
- Development and implement policies and procedures consistently as written
- Demonstrate the behavior that is expected of subordinates
- Demonstrate caring and courteousness and treat people with respect
- Demonstrate concern for employee safety and well-being

When trust permeates throughout an organization there is an opportunity for organizational growth, productivity, increased employee engagement, and relationship building and sustainment.

5.3.2.3 Lean Thinking Leaders

Lean thinking leaders are not only good for performance improvement but are also key to facilitating customer satisfaction. These leaders know and place high importance on the services they provide through ensuring that they understand the needs and desires of the customer. They demonstrate an uncanny knowledge of the services or products provided to their customers, and they continuously seek ways to improve their products and services. They use lean tools to improve performance and train organizational members on the various lean tools and how and when to use the different tools to continue to improve performance to meet and or exceed their customer expectations.

5.3.3 Tools that Facilitate Innovation

Innovation does not just happen; it takes work and a strategic focus on implementing elements that act as a facilitator to achieve innovation. These elements are shown in Figure 5.4 and discussed in more detail in sections 5.3.3.1 through 5.3.3.4. Innovative ideas are frequently sought after in high-performing organizations.

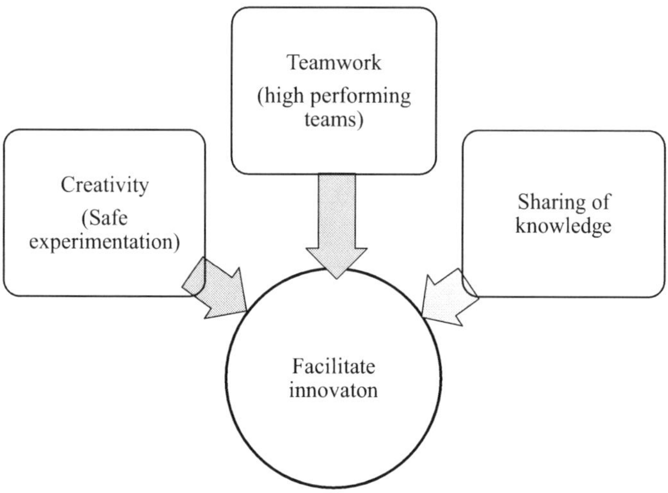

FIGURE 5.4 High-performance facilitate innovation

In a high-performing culture innovation can be facilitated through empowering employees to exert and exercise their creativity, working together as a cohesive team, and sharing their knowledge freely.

5.3.3.1 Teamwork

A group of people working together in a team setting sharing ideas and concepts provides a means to develop and discover ideas and concepts that can improve innovation. The collective brainstorming of effective teams is key in facilitating improvements to systems, discovering novel ways of approaching problems, and discovering new ways of accomplishing work and even new products.

Some benefits of working in teams include the following:

- Increase the possibility for innovation due to the sharing of ideas
- Facilitation of better problem solving and innovation through the sharing of ideas and knowledge
- Enhancement of team member personal growth through exposure to knowledge from other team members
- Expansion of creative ability of each team member through brainstorming and communication creative ideas and thoughts

There are many benefits of working in teams that are not listed in the list above. The point is that working in teams has the capability to facilitate productivity and improve company product offering and quality. Teamwork is not only good for the company's bottom line, it is also a growth mechanism for employees as they can gain from the knowledge sharing and networking that takes place in a teaming environment.

5.3.3.2 Creativity

Creative employees are able to collaborate effectively and share ideas that are designed to improve processes, the way work is performed or develop new products that can be added to the company's portfolio. These employees have the innate ability to work with other team members and transfer ideas into reality. Some benefits of encouraging creativity in the workplace include inspiring and motivating workers, promoting better and more efficient problem solving, and encouraging teamwork. In addition, creativity can be a facilitator of innovation and proactive issue identification and problem solving.

5.3.3.3 Sharing of Knowledge

The workplace of today places value on working in teams where feasible. Bringing together workers with different skill levels and experiences is the best way to seek solutions that are optimal. To facilitate knowledge sharing and collaboration some organizations used an open floor office arrangement where workers are given a laptop during onboarding and are not assigned an office or a desk (no walls to hinder communication). The office environment consists of a room with desks arranged in a manner so that workers can collaborate throughout the work shift. Collaboration is key for the exchange of ideas that can potentially create new processes, improve

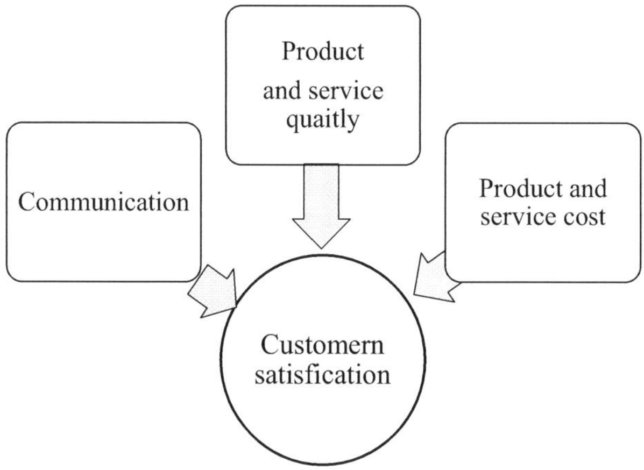

FIGURE 5.5 Components of improved customer satisfaction

efficiency of current processes and systems, and facilitate innovation. Sharing of knowledge between workers is a good way to develop employees and increase knowledge retention for an organization.

5.3.4 Customer Satisfaction

High-performance cultures are recognizably great for employees and employers; however, these cultures are also optimal for customers. Happy customers are repeat customers, and importunately they help attract other customers and serve as a great avenue for recommendations and advertisements of the service they receive. Customer satisfaction can be facilitated through communication, improving product quality, and increasing costs as shown in Figure 5.5.

5.3.4.1 Communication

Communication with customers is a must for a business to remain productive and relevant. Customer communication and relationships flourish in high-performing cultures. In high-performing organizations communication exchanges between management and workers are fluid, and this fluidity is carried on to their customers and can be seen in the way the organization functions. Communication with customers is a great way to keep abreast of their wants and needs. When referring to communication, all avenues of communication included should be specific written, verbal, and nonverbal such as body language.

Communication attributes in the workplace that can not only improve the work environment for employees but also impact customer relations across the board include:

- Relationship building
- Conflict resolution
- Builds trust

- Increase awareness of self, worker needs, and the needs of the customer
- Increase employee engagement.

There are many other benefits of effective communication that management should tap into to improve employee and customer relationships.

5.3.4.2 Product and Service Quality

Product or service quality is what customers are seeking when deciding to conduct business with an organization or purchase a product. The quality of the product provided to the customer whether it be services or supplies depends solely on those who have a responsibility to support the services or products. Therefore, everyone involved in the cycle must have a customer service mindset and focus on quality enhancement and continuous improvement. It has been proven time and time again that satisfied customers are not only repeat customers but can also help expand the customer base and thereby improve the bottom line for the company.

5.3.4.3 Product and Service Cost

The cost of products and services is always of concern to customers as it ultimately impacts their bottom line and profitability. In today's environment the cost to produce goods continues to increase along with the challenge of obtaining needed supplies at an acceptable frequency. The cost of living continues to increase at a pace where income received by workers is not keeping up with the pace. As such companies are having difficulty obtaining and retaining workers; therefore, in order to be competitive, they are offering more pay and bonuses to attract and retain workers. This cost is passed on to the customers creating affordability challenges. To balance this cost increase cycle, managers must become creative in improving process efficiencies and modify where feasible while not impacting quality or services. The primary cost focus is to ensure that the services and products provided are not significantly higher than competitors or a company can price itself out of the market and lose customers.

5.4 MAINTAINING A HIGH-PERFORMANCE CULTURE

Maintaining a culture that will facilitate high performance is not easy and will take time and attention from management. It will also require investments such as training, employee engagement activities, a communication campaign, and more. Building and maintaining a high-performing organization starts with knowing the current state of the culture. It is important for the management team to focus on facilitating a culture that can serve as the bedrock that underpins the way work will be performed. Sections 5.2.1.1 through 5.2.1.3 provide a road map that can be used to evaluate culture and determine the current state. Additional tools to facilitate culture evaluation are listed in Resource Packet found in Chapter 15.

5.4.1 Evaluating Organizational Culture

There are a number of methods that can be used to measure and evaluate cultural health within an organization. These methods can include reviewing documents such

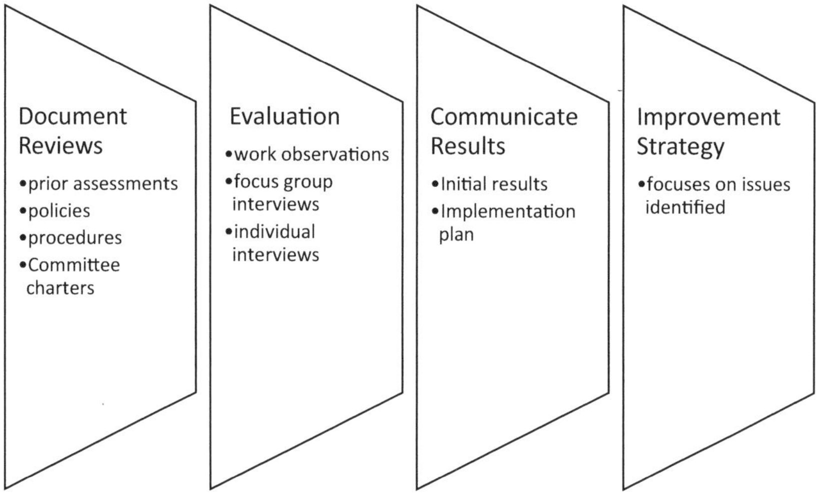

FIGURE 5.6 Culture evaluation steps

as policies and procedures, administering a survey, interviewing individual employees, conducting focus group discussions, and conducting work observations. Used alone these methods will not reveal the total health of a culture; however, collectively these methods can provide useful information that can be used to determine the health of culture and measures that can be taken to further improve culture.

The literature on culture states in general terms that culture can be created to display the creator's values, priorities, and vision. Therefore, it is valuable for leaders to evaluate the culture of their organizations at some frequency. Figure 5.6 outlines a process that can be used in evaluating culture.

Documentation Review – among the first steps in evaluating the culture of an organization is the review of documents such as applicable policies, procedures, charters, and previous assessment and culture evaluation results. In organizations where the policies and procedures are clearly documented and communicated workers are more likely to make every attempt to comply with the expectations set by these documentations. Workers can become confused and unsure of the appropriate behavior and expectations when there is a lack of documentation that clearly supports the vision and the culture of the organization. This ambiguity will have some impact on the culture and the compliance posture of an organization. Conducting a review of documentation as the first step of the process has the possibility to uncover important information that can inform a survey, focus group interviews, and individual interviews.

Evaluation – there are several mechanisms that can be used to conduct a comprehensive survey of their work culture. These methods include the use of survey questionnaires, documentation review, work observation, focus groups, and individual interviews. When using surveys to assess culture there is a need to ensure that the proper survey questionnaire is selected. There are many surveys or questionnaires on the market that are used to measure culture from various viewpoints. When using developed survey instruments, ensure that:

- The survey instrument contains the questions needed to provide answers that complement the purpose of the evaluation and the feedback being sought on the culture for the organization
- The survey has been validated
- The questions are clear and not written in a way that can be perceived as leading. Leading questions can invalidate the response received

Conducting focus groups and individual interviews is another effective way of gathering data to support a comprehensive and effective culture evaluation. A focus group should consist of a group of employees consisting of 6–12 participants. The reason for keeping the group small is to facilitate free flow of information exchanges. When utilized appropriately, focus group discussions are a valuable tool used to discover information on the opinions of employees. One should consider the following when conducting focus group discussions:

- Keep the list of questions short to no more than 15 questions
- Use terms that are familiar to the participants
- Use open-ended conversational type questions
- Do not use leading questions
- Each question should focus on a single element or activity
- Select participants randomly from a group that have knowledge of the topic
- Select a meeting location that would facilitate privacy for the group
- Select a facilitator that is trusted and objective and can shepherd the free flow of information

A focus group facilitator should be carefully selected as they are important in facilitating the exchange of information and ideas. The primary role of the facilitator is to guide and encourage group participation and discussion. Guidelines that can be used when selecting a facilitator are:

- Good communication skills
- Skilled in encouraging people to speak
- A good listener
- Able to control and guide the discussion if necessary
- Skilled in asking scripted predetermined questions and unscripted follow-on questions as needed
- Able to put the participant at ease
- Experienced in facilitating group discussions
- Able to tactfully deal with outspoken group members

Individual interviews can be used when there is a need to gather information privately from one individual or the interviewee is uncomfortable within a group setting. This method is an effective way to get feedback from senior managers. The same questions should be addressed for each focus group discussion. Additional questions may be developed for senior members of the leadership team. Individual interviews should be kept to the prescheduled timeframe. The use of individual and focus group interviews is a good mechanism to collect additional data on the cultural health of an organization as well as provide targeted areas of concern that can be used to inform the improvement strategy. The information obtained will provide insight into the views and opinions of the leadership team and the workers that directly impact organization culture.

Work observations provide an opportunity to allow the evaluator to observe work and see how work is actually being performed by workers. Observing work provides an indication of whether the policies and procedures are followed, and if it is acceptable for work to be paused if there is a lack of understanding of procedure, if regulatory noncompliance exists, or a safety hazard that needs to be addressed. The actions of workers during work performance can provide clues on the workplace culture. Some actions to pay attention to during work performance include the following:

- Was a pre-job briefing performed before work began?
- Did the workers performing the job participate in the pre-job briefing?
- Was the scope of the work discussed and followed?
- Did the worker perform the work according to procedure, work documentation, etc.?
- Were safe work practices used?
- Were the tools needed to perform work on the job site and available to the workers?

*Communication_*– When embarking upon the culture evaluation process, it is paramount that a communication plan be developed at the onset and followed. Communicating with workers the need to perform a culture assessment and the process that will be used to conduct the data collection is important to begin the process of obtaining buy-in. Once the culture assessment is completed a comprehensive communication campaign should be encountered to provide employees with the result of the assessment and management's commitment to address any weakness in the organization culture. The improvement strategy is communicated to workers as soon as it is developed and approved by management outlining the improvement areas and the process for addressing areas of concern.

Improvement Strategy – A comprehensive culture improvement strategy should be completed once the data obtained from the culture assessment has been analyzed and areas of improvement have been identified. The plan should address all areas of weakness identified. It is not necessary to address all issues at the same time; however, all issues should be included in a schedule with a time frame for addressing them. Metrics should be developed to track the plan's progress and made available for all employees to view.

5.4.2 Employee Engagement

The engagement of employees is key for an organization to gain access to the knowledge and skills they possess. It is also required for leaders to be able to influence employees to buy-in, participate, and embrace the work culture. Employee engagement is necessary to gain buy-in from employees. Buy-in from employees should be sought after with the realization that not every employee will buy into and support every initiative or decision made on behalf of the organization. When employees accept the decisions of leadership, they are more likely to go the extra mile to accomplish work and become active participants in the organization culture. Employee engagement has many recognizable benefits for organizations to include:

- Increased productivity and the desire to assist the organization to achieve at their highest level and meet the needs of their customers
- Increased facilitation of employee growth
- Empowers employees to participate in the business
- Increased development and maintaining trust
- Facilitate employee retention
- Increased creative and strategic thinking
- Better decision making that is in alignment with the goal and vision of the organization

In order for engagement to be highly effective as a strategy, not only must employees be engaged in their organizations, but they must also believe that they are included as an integral entity in helping achieve goals, objectives, and the mission.

5.4.3 Leadership Role and Engagement

Leaders play important roles in all activities within an organization from strategy development, policy enactment, championing change, process improvement, and facilitating culture. These roles should not be understated or underestimated as they form the basis of the culture of the organization as well as the path an organization takes to success and their ability to achieve their peak performance.[2] Leadership is about having the ability to get results from people that supports the value of the organization, mission, and business strategies.

The implementation of the mission and goal of an organization is an important responsibility of leadership. In fact, it is a primary responsibility for leadership. In order to accomplish this role, a leader must be committed to the organization and the vision. Leadership commitment is verbally communicated and modeled, documented in company policies and procedures, with the appropriate accompanying actions. Many employees will be more suspectable and become committed to the organization if they believe that the leadership team is committed. A relevant question is how can a leader demonstrate their commitment to the organization. Some actions that demonstrate leadership commitment include the following:

- They demonstrate genuine concerns for employees
- They develop and implement policies and procedures that are in the best interest of the business and the employees, taking into consideration the impact of those policies and procedures on the workers and their ability to perform

- They walk the talk. Leadership follows the policies and procedures that guide the culture and the way work is performed by the organization
- Conduct themselves in a manner that they are considered trustworthy
- Lead with honesty and humility
- They display respect for others in their actions and transactions

Leaders are generally looked up to by workers, and workers will typically emulate the actions of leaders. As such, leaders have the ability to shape an organization through their prevailing behaviors. Therefore, leaders should be cognizant that they are impacting the culture of their organization with everything they do.

5.5 DEVELOPING AND IMPLEMENTING A CULTURE IMPROVEMENT PLAN

Once the data has been analyzed and areas for improvement have been identified, it is time to create the Culture Improvement Plan. The improvement plan should address the areas that were determined to be below the desired outcome as well as address ways to maintain or improve the areas that were identified as strengths. The improvement plan should provide a detailed road map on how the program, process, and behaviors that are in need of improvement will be addressed. Also included in the plan are measures that will be taken to avoid eroding the areas that were identified as strengths.

REFERENCES

1. Frances, Alston. *Culture and Trust in Technology-Driven Organizations*, CRC Press Taylor & Francis Group, 2014.
2. Allen, Patricia, Alston, Frances, and Dekerchove Emily. *Peak Performance How to Achieve and Sustain Operations Management*, CRC Press Taylor and Francis Group, 2019.

6 The Art of Effective Communication

6.1 COMMUNICATION OF HAZARD ANALYSIS AND RESULTANT CHANGES

Have you ever engaged in a communication exchange and the message others received from that exchange is not what you believed you communicated? This scenario is not unusual and has been a reality for many during communication of critical information. Even the best communicators at times can experience problems in getting their message across to a diverse population of people. A reason for this conflict in what is communicated and what is heard is complicated and based on several factors such as the way communication was delivered, the choice of words, the environment, the body language including facial expressions of the communicator, and the list can go on. Communication within the business environment can present many challenges that go beyond the words that are spoken. I would contend that facial expression is one of the most important factors to be aware of coupled with other forms of body language. For example, some emotions that can be seen through facial expressions include:

- Happiness
- Frustration
- Confidence
- Confusion
- Anger
- Surprise
- Fear
- Sadness

The above attributes, when visible in facial expressions can have a significant impact when communicating workplace information such as safety and health program changes, injury and illness information, and policy changes. In fact, one's facial expressions can be a determining factor for many listeners as to whether the individual and the message can be trusted.

The exchange of information that represents the communication process is how people exchange ideas and accomplish tasks and activities. Communication can take the form of verbal, body language, and eye contact all during the same exchange. A broad definition of communication is that it is a process of sending and receiving messages through verbal or nonverbal means. Also included in that definition

DOI: 10.1201/9781032624389-8

TABLE 6.1
Strong Communicator Best Practices[2]

Best Practice	Value Proposition
Truthful in all communication	Truth telling facilitates trust which becomes the foundation for productive interactions
Good listener	To hear and understand the viewpoint of another person
Choose precise words to deliver the message	Careful communication eliminating provoking words such as disapprove, unreasonable, and unfair
Pay attention to the emotions displayed	Understand and pay attention to how the lister's mindset may influence reception of the message being delivered. Select the appropriate emotional setting to deliver information
Ability to read body language	Words do not portray the entry story. The tone of your voice, facial expressions, eye contact, posture, gestures, etc. provides feedback on how the message is received
Understand the importance and role of timing	Know when to deliver the message to facilitate acceptance
Seek opportunities to prevent mind fields	Pushy communication can turn off others and cause them to tune out or misunderstand your message. In some cases, the listener may launch out with derogatory words

are writing and graphical representations (for example, maps and charts), body language, and behavior.

Workplace communication forms the basis of developing trusting relationships that encourage employee engagement and trust among workers and management. Open honest and effective communication is important in improving cultures and enhancing workplace safety and health. Table 6.1 lists some best practices to engage in during communication. These best practices have the capability of increasing influence and acceptance of the message being delivered.

6.2 ARE YOU A GOOD COMMUNICATOR?

Communicating with workers from all walks of life, technical professionals, non-technical professionals, and leaders requires one having the ability to modify their communication skills as needed to get the message across. When asked, most people will affirm that they believe that they are an average or good communicator. One important characteristic of a good communicator is being aware of your actions and words during communication. Figure 6.1 and Table 6.2 list additional important attributes of a good communicator along with important benefits. The Figure and the Table do not contain all elements necessary to be considered a good communicator; however, they consist of many of the basics. Note the two elements that stand out as they are considered critical elements of communicating effectively.

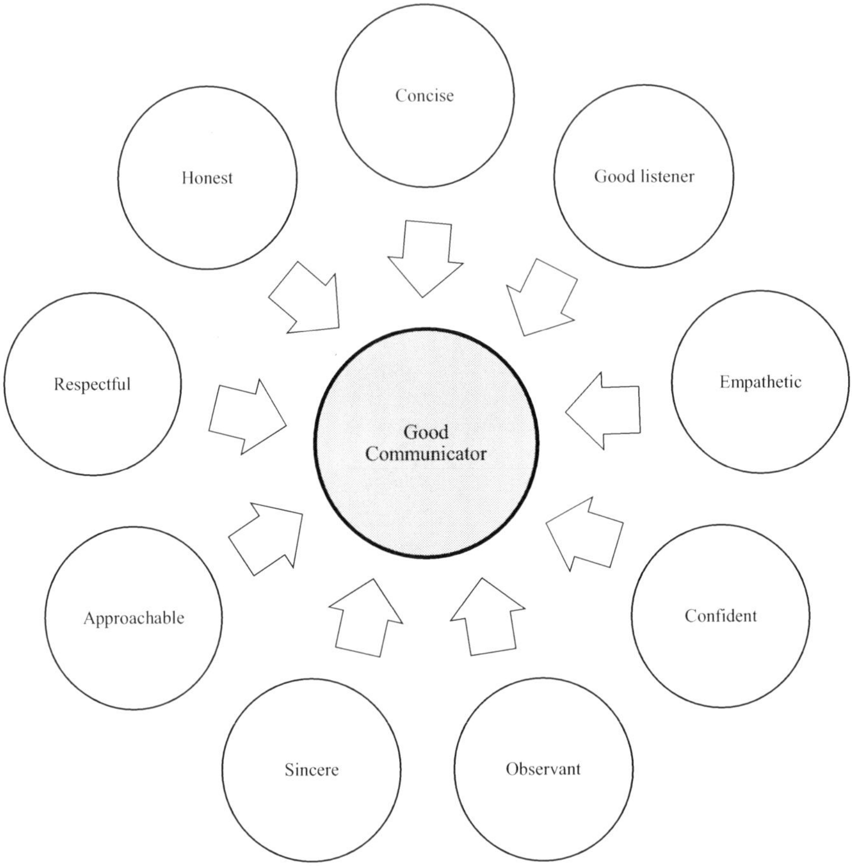

FIGURE 6.1 Good communicator characteristics

For some, good communication skills and the use of the characteristics discussed come easy, and for others it is difficult to make use of these characteristics when interacting and engaging in dialogue with others. Providing feedback on safety and health issues, monitoring data, or health hazards directions requires the use of skills that will get and hold the attention of the audience in a manner that will ensure the message is received, understood, and accepted.

6.3 DATA REPORTING AND COMMUNICATION

Communication is a large part of safety and health professionals' job responsibilities. Safety and health professionals collect a large amount of data in today's increasingly advanced digital age in their plight to quantify exposures and reduce risk for employees while in the workplace and prevent injuries and illness. The data collected is through various means using different media and equipment. Assimilation

TABLE 6.2
Good Listener Characteristics and Benefits

Characteristics	Benefits
Honesty	A key characteristic of good communicators is honesty. Having a reputation of being honest increases trust and the message communicated will be believed and embraced
Respectful	Showing respect during communication through being attentive and polite. Also included in showing respect is actively listening to others and what is being communicated
Good listener	Effective communication requires an exchange of ideas between two or more parties. Listening during the exchange of ideas and thoughts is essential in understanding the topic and building relationships. Attentively listening to others demonstrate your respect for them and interest in the topic being discussed
Concise	Direct clear communication will be key in conveying message intent. Being concise requires an awareness of the time spent communicating the message yet not rushing to conclude the discussion or delivery of message
Sincere	Being genuine during communication can aid in trust building, and it allows safe sharing of feeling that can connect those that are engaged in conversation
Empathetic	Showing empathy can assist in building trusting relationships. Empathy provides the means to understand the plight and viewpoint of others. In addition, it demonstrates to others that you care for their well-being
Approachable	Speaking in a tone and exhibiting body language that demonstrates approachability encourages others to engage in communication
Confident	Communicating with confidence opens the door to credibility and believability. When displayed confidence can help get the message across in a manner where the communicator is not questioned regarding their authenticity
Observant	Capable of observing nonverbal communication is important to gaining insight into the feeling of the party in which one is engaged in information exchange

of the various streams of data is essential for decision-making and evaluation of program performance, projects, and policies. When communicated appropriately, data is an effective way to facilitate and inspire action. Figure 6.2 outlines considerations for communicating messages effectively to a diverse audience or individuals.

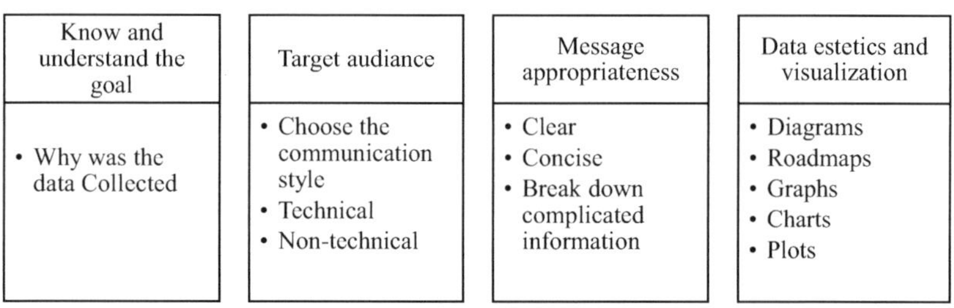

FIGURE 6.2 Data communication channels

Managers are seeking to make use of data more in their decision-making process to assist them in reaching their goals. Using data effectively presents a learning curve for companies and their employees as they learn how to collect, use, and communicate data in a way that will educate, enhance performance, and inform decisions.[1]

6.3.1 Communication with Employees

Safety and health (S&H) professionals and management continuously communicate with the workforce even when they are not sending messages verbally. Workers constantly observe the actions of management and these professionals. Therefore, it is incumbent that they walk the talk and follow policies and procedures that are in place to conduct business and maintain a healthful and safe environment. It is important that the S&H professionals be viewed by workers as an integral and respected part of their team. Having respect for workers is a significant factor in bridging the communication gap and enhancing trust that the S&H professionals will work on their behalf to enhance their safety and health while at work. In order to gain and retain the respect and trust of workers, safety and health professionals must be viewed as trustworthy displaying at all times the following attributes during interactions. More detail on these attributes of trust is discussed in detail in Alston (2017).[2] The attributes of trust not only distinguish good leaders from marginal leaders, but these attributes are also beneficial in building relationships and enhancing communication. These attributes are:

- Concern for employees
- Open and honest communication
- Viewed as being competent in their roles
- Reliable and dependable in responding appropriately
- Building and maintaining relationships with the workforce
- Identification with the employees in the work environment having comparable values working toward the same goal

S&H professionals are considered leaders in the environment in which they work. These attributes of trust serve as the foundation for effective communication between people in organizations. When communicating with workers, the tone of voice is significant especially when delivering information that is urgent or information that is non-urgent but informative. The tone of voice can enhance or degrade communication and impact the way the message is received or understood by listeners. When referring to the tone of voice, reference is being made to the pitch, volume, rate of speech, and emphasis points.

Another key aspect of communication with workers includes the body language displayed because different workers can interpret the various types of body language (facial expressions, body movement, eyes) differently. The body language meanings are read differently depending on the culture of the individual. For example, in some cultures direct eye contact for a period of time is expected during communication and may be inappropriate or uncomfortable in another culture. Because of the differing interpretations of body language, it is advisable to be cognizant and minimize the use of body movements.

6.3.2 Communication with Supervisors

This section on communicating with supervisors is designed to bring to the forefront that communicating with supervisors at times can be different from communicating with workers or senior leaders. Supervisors are close to the work that is performed in their environment and the workers performing various tasks to accomplish the goals of their group or department. Therefore, some supervisors may feel that the challenge of ensuring the well-being of the worker and accomplishing work may conflict especially when there is a pressure to place production above safety.

A supervisor cannot be successful in their roles if they are unable to foster trust and build relationships with their subordinates. However, recognizing that there is a balance and a line between supervision and workers, the workers must believe that their supervisor or direct manager is supportive and is concerned for their well-being. The relationship between a worker and their supervisor forms the basis of communication and the reception of the various types of messages delivered. It may be helpful when communication with supervisors to take into consideration the worker–supervisor relationship and ensure that the information is tailored to the worker with the expectation that the supervisor will pass the information along to the worker. Supervisors can be good partners when delivering messages on workplace risks and mitigation actions.

6.3.3 Communication with Senior Leaders

Communication with senior leaders can take on a different form when discussing the safety and health program and work environment. These leaders will mostly be concerned with the program at a high level, for example, is the program effective in protecting workers and enabling safe performance of work and productivity. They will also be concerned with the financial investments and impact of program implementation. When communicating with these leaders focus on providing the high points of program performance such as safety statistics to include injury and illness along with preventive measures instituted to prevent recurrence. Just as with other programs, the safety and health programs are in competition for resources; therefore, when requesting support, it is advisable to provide justification that includes cost, benefits, and expected outcomes. However, leaders are apt to provide the resources needed to ensure a comprehensive program that will preserve the health and safety of the workforce.

6.3.4 Communication with the Occupational Medicine Provider

Communication with the medical community is an action that most people are accustomed to while engaging in wellness visits and visits during specific illnesses.

Therefore, many employees are comfortable speaking with a doctor or a nurse and for the most part trust them to deliver a message regarding their health without biases.

The occupational medicine physician is a credentialed medical doctor that specializes in the area of occupational health. These specialized physicians focus on managing the health and safety and wellness of workers while in the workplace. The occupational physician works closely with the health and safety staff, especially the industrial hygienist in preventing and treating work-related injuries and illness and support promoting the well-being of employees; developing and implementing workplace health and safety policies and procedures that are supportive of employer goals and protect workers.

Responsibilities of an occupational health physician can include:

- Physical Examination: Performing pre-employment and post-employment physical examinations
- Policy Development: Collaborating with company representatives and safety and health professionals in policy and procedure development to improve and maintain safety and health, reduce absenteeism
- Medical Surveillance: Develop and implement medical surveillance programs (dictated by regulations or best practices)
- Health Screenings: Performing screenings to identify underlying conditions that could affect an employee's ability to work safely such as drug and alcohol screening and human reliability programs
- Employee Fitness: Assess employee's fitness for duty. Evaluating employee injuries and determining their severity
- Medical Consultation: Provide consultation to individual workers on health conditions and disabilities
- Hazard Identification: Working with employers and employees to identify and mitigate potential health and safety hazards
- Vaccinations: Offering vaccinations to workers if the risk of infection in the workplace is greater than in the general population
- Health Promotions: Advising employees on how to improve their well-being and fitness to prevent work-related injuries
- Worker's Compensation: Reviewing patient's health and working conditions in support of worker's compensation claims
- Public Health Advance: Advising employers and employees on public health and work safety issues

Some of these responsibilities are predicated upon whether the occupational physician works directly for a company or works in a clinic that serves employees from a variety of workplaces. Occupational medicine physicians play a critical role in the health and safety of workers while in the workplace.

6.3.4.1 Workforce Health Monitoring

An important aspect of ensuring a workplace for workers that is reasonably safe involves having an effective safety and health program and processes that collect exposure data that is analyzed to determine the state of the work environment. Today in the work environment various means are utilized to collect data such as:

- Benchmarking programs that are called out as being effective
- Collection and evaluation of exposure monitoring data
- Review of internal assessment reports and associated data
- Review of external assessment reports and associated data

An important aspect of communication within the work environment is communicating safety and health information to workers and management in a manner that the information is easily understood. It is not easy communicating exposure monitoring data, and the complexity of communication can result from factors such as the results of the monitoring and the knowledge of the population or person in which the data is being communicated. Regardless of the competing factors, it is incumbent that the communicator possesses the knowledge and skills to deliver the information in a manner that can be understood and reasonably accepted. For example, an overexposure may not have occurred. However, the air monitoring results indicated asbestos above the action level while removing insulation containing asbestos for two hours while wearing respiratory protection. Depending on the knowledge of the worker in understanding occupational exposure limits, action level, and the protection factor (PF) for the respiratory protection worn, there may be a challenge in providing an explanation that is acceptable to the worker. The message to the worker may include that their exposure is considered to be a protective exposure meaning that the protection factor of the respiratory protection device should have prevented them from experiencing a harmful inhalation hazard. The employee may focus on the fact that the monitoring result is above some level and not understanding the meaning and purpose of action level when assigned to chemicals. In such a case, the explanation and communication event may take longer, and it may be necessary to engage other professionals or the occupational medical physician in the discussion. The occupational medical physician can address the aspects of the monitoring results as it pertains to potential health impacts. Having an explanation of health impact from a physician may provide ease of concern and help the worker gain an understanding of the monitoring results and potential health impacts.

Another aspect that safety and health professionals need to be cognizant of when communicating the exposure result in the example is the tone of voice used in communication and body language as well as the words used during communication. Examples of the potential impact of these attributes when communicating exposure monitoring data to workers.

- Tone of Voice – is key in enhancing the effectiveness of the information being communicated and can impact how the information pertaining to the monitoring results is received and understood by workers
- Body Language – can send an unwanted message that a problem exists or the person delivering the message is withholding information

a. Facial Expression – is a part of delivering the message because your facial expressing can portray emotions such as confusion that the worker may observe and may lend itself to a lack of disbelief in the information being communicated and a lack of track for the safety and health professional
b. Gestures – can contradict the message that the safety and health professional is trying to convey to workers. For example, too much hand movement may be distracting to the worker, and the essential information may be overlooked.

- Written Communication – when communicating results are written such as an employee exposure report, care must be given to the clarity of the message. The message must be written taking into consideration the knowledge level of the workers.

6.4 TYPES OF SCENARIOS COMMUNICATED BY SAFETY AND HEALTH PROFESSIONALS

The health and safety professionals are charged with communicating with employees, supervisors, and management daily. Admittedly, the nature of these communications is not always considered good news, can impact the performance and at times the speed at which work is accomplished, and may require some specific actions from management that can impact financial allocation, for example, repairing an equipment that is producing loud noise above the regulatory limit. To demonstrate the potential types of communication, we have offered a couple of scenarios that can frequently happen in the work environment and provide some feedback on communication channels and potential ramifications.

6.5 POTENTIAL OVEREXPOSURE BASED ON MONITORING DATA

Let's consider how one would communicate a potential overexposure to an employee supported by exposure monitoring data. The scenario for consideration is that an employee is working in an environment where the facility is old constructed in the 1920s and there are various hazards present as a result of equipment painted with lead paint that is currently used, however infrequent. A critical piece of equipment went offline and required maintenance to repair and get back online. A team consisting of workers, industrial hygiene, safety, engineering, and work management personnel walked down the project and discussed the procedure that will be used to safely perform the work. The work was planned and scheduled.

While performing the work, all work procedures were followed; the two employees performing the work task wore the prescribed protective equipment (PPE) that included an air supplied hood, Tyvek, and gloves, and the two maintenance workers were fitted with a personal monitoring device to collect breathing zone samples. The industrial hygienist made the decision to monitor both workers because it was determined that they would not always be in the same vicinity of the work area during the entire work evolution. Once the job was completed in the work area the sampling pumps were collected, and the sample medias were submitted to an American Industrial Hygiene Association (AIHA) accredited laboratory for analysis.

The results received from the AIHA laboratory for one employee indicated that the eight-hour exposure result was above the occupational exposure level (OEL). The results for one employee were below the OEL. The communication challenge for the industrial hygienist in this scenario becomes one of four key discussions:

1. Communicating to the two workers that their exposure results were different
2. Communicating with the worker with the sample results above the OEL
3. Communication to the worker with the sample results below the OEL
4. Communicating that the results received although one above the OEL is not expected to impact the worker's health

Industrial hygienists should prepare wisely before engaging in a conversation to deliver the sampling results. However, the discussion should happen soon after the results are received from the laboratory. Some preparatory actions may include the following:

- Discussion with an occupational medical provider
- Verify that the sample was collected properly with calibrated sampling pumps using the appropriated sampling media
- Discussion with the laboratory that provided the results to ensure that the sample was analyzed properly, and the results reported were accurate
- Anticipate and be prepared to respond to potential questions that may be posed by workers

It is advisable that the industrial hygienist have a discussion with the medical provider to gain information or suggestions on the most effective means to discuss the concept of protected exposure. The workers wore respiratory protection devices that should have precluded them from being exposed internally because of the protection factor afforded to them by wearing the respiratory protection device. The occupational health provider will be able to explain in terms related to health impact of the exposure. It may be advisable to request the presence of the occupational medical provider when discussing the results of the monitoring for each employee. The discussion with each employee is a private discussion since personal exposure data is considered Personally Identifiable Information (PII).

An important attribute of collecting samples is to ensure that the sampling equipment is in the appropriate condition, the proper sampling media is used along with flow rates and handling of the samples once collected. The sampling pump must be maintained and calibrated for the specific chemical constituent in this case lead. As a part of ensuring that the pump is in proper working condition there needs to be a focus on the entire sampling train to include the sampling media. Verifying the sampling protocol, equipment, and implementation of a sound sampling collection procedure will be helpful in communicating with confidence the results of the sampling evolution. For example, if the desire is to have the sample analyzed using the NIOSH Method 7082 for lead, then the samples must be collected based on criteria set by that method. Specific criteria of that method include collecting on a 0.8-μm cellulose ester membrane filter, a flow rate of 1–4 L/min, and volume between 200 L @ 0.05 mg/m3 and MAX: 1500 L,[3] therefore, enhancing confidence in the results reported by the AIHA laboratory.

Before discussing the data results the industrial hygiene will want to ensure that the data is accurate, especially since one monitoring device data indicated overexposure and one didn't. It is not often that the need arises to question the laboratory on the results submitted; however, in situations such as the one presented it is a good idea to go the extra step. Having detailed knowledge of the methods used for analyzing the samples and the handling methods can provide additional insight into the results and provide additional information that can be used when discussing the results and answering potential questions.

Another issue to consider prior to communicating the results of the sample is to anticipate some of the potential questions that may be posed by the workers. Prior questions posed by workers can be used to anticipate questions and prepare for responding to questions regarding exposure potential, data collection, accuracy of results, health effects, etc. Brainstorming with colleagues is another tool that can provide insight into questions based on their knowledge and experiences.

The results of the samples collected should be communicated to each worker individually. The employees should be provided with a written copy of their exposure report signed by the industrial hygienist and/or the industrial hygiene manager. An example of the completed exposure results communication is shown in Figures 6.3 and 6.4. The results reported in the two figures, although representing different sample outcomes, are reported in the same manner and state that facts of the results are as known. The sampling reports should be used when communicating results and the conversation should proceed from that point taking into consideration the questions posed by the worker. During discussion ensure that the results are presented in a manner that can be understood by the employee. There may be a need to limit or avoid using highly technical terms and maintain eye contact while controlling facial expressions. Body language and tone of voice can make or break the delivery of the message in this instance when attempting to communicate that possible negative health impact is not probable and that the exposure is considered to be protective because of the respiratory protection device worn during the job and the associated protection factor (PF). The results should be presented with clarity, disclosing all the information known concerning the sampling campaign and laboratory results. Questions should be answered with honesty displaying openness to answering all questions. Avoid the following while communicating:

- Rolling of the eyes
- Displaying impatience because the worker is asking a lot of questions
- Closed and unwelcoming body language
- Harsh or condescending tone of voice
- Withholding information
- Dismissal of worker point of view
- Sighing

Notification of Personal Monitoring Results

Name: Employee #1 Date: January XX, 20XX

Location: Building 2424, Room 1021

Contaminant Monitored: Lead

Summary of Task Performed: Realignment of power press through removal of old shaft and replace with a newer mod Plots•Charts•Graphs•Roadmaps•Dgrams•visualizationData estetics and informationcomplicated Break down •Cncise•Clear•appropriatenessMessage technical-Non•Technical•stylecommunication Coose the •Target audiancedata CollectedWhy was the •goalunderstand the Know and el

Results: 55 ug/m^3 OEL: 50 ug/m^3

Summary of the Results: The personal sample result was above the OEL

Recommendations: None

Sampling Results Reviewed and Validated by: John J. XXX

Signature:

FIGURE 6.3 Worker #1 exposure written communication

The actions above say to the workers that you are not concerned with their thoughts and feelings therefore may not be concerned with their health and safety. Furthermore, the worker may choose not to believe the data you presented or your explanation of the data.

The difference in the message to the two workers requires the communicator to understand the differing ways of communicating exposure results recognizing that the reactions to the information by the two workers may be different. Therefore, the types of questions and time required to explain the monitoring results will be different. The worker with elevated exposure results will more than likely have more questions and may be concerned about future potential health impacts.

6.6 I SMELL SOMETHING IN MY WORK AREA

Another scenario that can be challenging for safety and health professionals to communicate with workers is when a worker or workers report that they smell something

Notification of Personal Monitoring Results

Name: Employee #1 Date: January XX, 20XX

Location: Building 2424, Room 1021

Contaminant Monitored: Lead

Summary of Task Performed: Realignment of power press through removal of old shaft and replace with a newer model

Results: 22 ug/m^3 OEL: 50 ug/m^3

Summary of the Results: The personal sample result was well below the OEL

Recommendations: None

Sampling Results Reviewed and Validated by: John J. XXX

Signature:

FIGURE 6.4 Worker #2 exposure written communication

in the work area that is not noticed by others and the smell according to them is infrequently noticed. Once alerted of the concern the industrial hygienist will begin evaluating the work area to determine if the source of the issue can be identified and potential exposures can be quantified. Often in this scenario the result of the evaluation may yield inconclusive results, and the data obtained does not indicate an exposure potential. The challenge of explaining to workers that the results of the assessment did not indicate an exposure issue and that the smell had not been identified.

Communication becomes even more challenging once the workers become convinced that they have been over-exposed to something and as a result they are experiencing symptoms that they believe are associated with exposure. It can be frustrating for industrial hygienists to spend the time necessary to ensure that the workers have all of the information that they have with regard to the issue and the path taken to discover and remedy the source of the smell. Regardless of the beliefs of the worker, communication should take a path of openness and honesty.

REFERENCES

1. Booher, D. *Communicate Like a Leader: Connecting Strategically to Coach, Inspire, and Get Things Done*, Berrett-Koehler Publishers Incorporated, 2017.
2. Alston, F. *Lean Implementation Applications and Hidden Costs*, CRC Press Taylor & Francis Group, 2017.
3. LEAD by Flame AAS, https://www.cdc.gov/niosh/docs/2003-154/pdfs/7082.pdf

Part III

Chapters 7–11 focus on identification, mitigation, and control of hazards found in the workplace.

DOI: 10.1201/9781032624389-9

7 Hazard Analysis and Documentation

7.1 WHY CONDUCT HAZARD ANALYSIS

A safe workplace is value added for businesses, employees, and customers. Unsafe workplaces can create challenges for a business such as:

- Employee Absence: Injured employees having to take time away from work to recuperate
- Accidents: Slips, trips, and falls are a major threat to employees, and OSHA estimates that they account for more than a third of fatal workplace injuries
- Fatality: Potential death of workers leaving families in turmoil
- Notoriety: Damage reputation of the company as a place that is unsafe to work
- Loss of Income: Potentially loss of businesses and clients
- Increased Premiums: Increased insurance premiums can result from reported and documented unsafe work conditions
- Increased Lawsuits: Increased litigation could result from employee injury as a result of an unsafe work condition
- Health Issues: Poor ergonomics can lead to health issues like carpal tunnel syndrome, sore back, and tendonitis
- Regulatory Compliance: Failure to comply with regulations can leave a business open to significant risk

Taking into consideration the list of challenges presented, it is clear that safety and health can add value to a business, the work environment, and the livelihood of workers. Therefore, preventing workplace injuries and illness is a valuable proposition that is being embraced more by organizations. Injury prevention can occur by reviewing and analyzing work processes, establishing procedures on safe job performance, and ensuring that employees are knowledgeable and trained in performance of job tasks. One of the best methods to use to determine and establish proper work procedures is to conduct a job hazard analysis (JHA). The issues identified in the job hazard analysis can be controlled or eliminated and prevent workplace hazard from occurring thereby preventing injuries, illness, or even death of workers. Deployment of an effective job hazard analysis process has the potential to result in fewer worker injuries and illnesses, safer and more effective work methods, and reduced workers' compensation costs.[1]

DOI: 10.1201/9781032624389-10

7.2 JOB HAZARD ANALYSIS (JHA)

In order to understand what a job hazard analysis is, several concepts must first be defined.

> ***A job is any physical or mental task that has been given as a responsibility to an employee and bears a positive or negative consequence based on job performance.***
>
> ***A hazard is something that can potentially cause harm. A hazard could be a situation or action that leads to injury or illness if not adequately controlled.***
>
> ***A job hazard analysis (JHA) is a strategy for identifying hazards before they arise. It focuses on the worker's connection with the task, the tools, and the work environment.***

After identifying uncontrolled hazards, the employer should ideally make efforts to eliminate or decrease them to an acceptable risk level. The JHA is usually performed when there has not been an initial analysis, when there is a new job task, when an incident or accident occurs, when a change has occurred in the performance of a previous task in which a JHA has been performed, and when employees have concerns about a job task. A variety of terms have been used when referring to a JHA, including Pre-Task Hazard Analysis, Job Task Analysis, Pre-Job Plans, Job Safety Analysis, and Pre-Task Planning and Safety Task Analysis.

JHA has been utilized most frequently in the construction industry, followed by process industries and healthcare settings. JHA's primary objective is hazard identification, but it has also been used for other purposes.[2] Other objectives of the JHA include to determine the training, equipment, and monitoring that are required to properly execute job duties in line with current work best practices and regulations. For each work task, the JHA identifies health risks and specifies mitigation actions to enhance safety of performing work tasks. Personal protective equipment (PPE) such as eye and face protection, head protection, foot and leg protection, hand and arm protection, body protection, and hearing protection may be prescribed based on the result of the JHA. Some of the factors that determine what task would be subjected to performing a JHA include frequency and severity of workplace accidents, task potential for injuries, new work processes, and non-routine work.

According to the Occupational Safety and Health Administration every day in the United States, numerous workers are killed or injured at work. The majority of debilitating workplace injuries usually occur during the first few weeks of an individual's employment primarily resulting from poor safety training of workers, unsafe work procedures, inadequate personal protective equipment, and no safety oversight or supervision of workers. These work injuries are usually because new workers are often unable to identify workplace hazards. The OSHA requires employers to provide a place of employment for workers that is free from recognized hazards that are likely to cause injury or death. The most common and useful tool used to comply with this requirement is the JHA. Many OSHA standards require hazard analysis such as analysis for PPE selection, lockout tagout, hazard communication, emergency action plan, etc.

The Basic Stages of a Job Hazard Analysis:

- Selecting the job to be analyzed
- Breaking the job down into a sequence of steps
- Identifying potential hazards
- Determining control measures to overcome these hazards.

Job Selection: A JHA can be completed for every task on a work site, whether it is routine or non-routine. Even occupations that just do a single activity should be evaluated by assessing the work environment. In general, JHAs for work with the highest frequency and severity of injuries, or the largest potential for accidents, should be completed first. JHAs should be done for work with new processes and procedures or where processes have changed and for infrequently performed work.

Task Breakdown: Breaking down work activities into individual tasks may appear to be a time-consuming process, but it is necessary for accurately appraising the hazards connected with the job. JHA software may be used, and it makes this step considerably easier by importing a library of past evaluated tasks stored that can be transported to the current task being analyzed.

Hazard Identification: Identify any hazards that may be present at any given time during each task. The hazard may not necessarily be found in the task itself but within the area where the activity is performed. When determining hazards, assess the risk of injury each hazard poses. Risk matrix calculations of the probability and severity of the hazard may be used to determine overall risk. JHA software is helpful in automatically calculating risk through defined levels of probability and severity.

Appropriate Controls: The final phase in a JHA is to identify measures that could be implemented to prevent recognized hazards from causing possible injury or sickness. If possible, a hierarchy of controls and personal protective equipment (PPE) should be implemented. If hazards cannot be eliminated, alter the work process, and reduce the risk to the greatest extent practicable. This includes using resources such as a Safety Data Sheet (SDS) for a chemical hazard which provides information to assist in understanding and mitigating the potential hazards. Also gather input from line supervisors and employees who may have identified missing controls or dangers during the hazard analysis process. JHA software can make this important process easier for everyone involved. Some software can help track potential hazards in the workplace and calculate the risks of each hazard automatically.

Job Hazard Analysis is an important component of safety management systems and is often conducted manually, relying on the safety and health personnel's experience and observations as new or inexperienced workers may not have the capability of identifying all the hazards in the work environment. After the initial analysis, JHA are only effective if they are reviewed and updated on a prescribed periodicity. The JHA team can comprise Safety and Health Manager, Safety and Health Team members, employees, supervisors, and Engineers/Technical Advisors.

Before beginning a JHA for a specific task, it is suggested to take the following actions:

- Employee Involvement: Front-line workers are the ones who understand their jobs best, and their knowledge, experience, and perception are helpful in identifying work hazards.
- Reviewing Accident Records: Examine any records of accidents, injuries, occupational illnesses, and equipment or mechanical failures at work.
- Conduct a workplace survey: After questioning your employees about any hazards in their workstations, discuss ways to eliminate or control them.
- Prioritize Work Tasks: JHAs should be undertaken first for jobs with the highest risk.
- Describe the Work Task: Observe the employee complete the task and examine the steps that make up the task. Taking images or videos of the employee performing work is useful in determining work activities and worker behavior.

JHA can be performed by the JHA team through observation or discussion. The team could directly observe job tasks to evaluate each job step and understand the hazards involved. This method may create a distraction if questions are asked during the performance of work. Observing work without an analysis is not the most preferred method since hazard identification and mitigation has not taken place. Pertinent information can be gleaned by asking workers questions and making use of their understanding of current practices, issues, and challenging areas.

JHA undertaking through discussion requires employees with knowledge of the job to gather and discuss a specific job activity step by step. Job analysis is not frequently performed on the job site. Some details may be unnoticed when the discussion is not done at the job site.

Performing a JHA has several advantages, including the ability to set performance criteria, standardize processes based on recognized safe practices, and prescribe PPE when necessary. It also helps with ensuring compliance of safety and health regulatory agencies such as OSHA regulations, reduces injuries, absenteeism resulting from employee injuries, increases productivity, protects personnel health, and boosts morale.[3] JHA is a vital technique for effective safety and health management. It will improve an organizational capacity to create an inventory or portfolio of hazards and risks associated with various jobs, job steps, and particular activities done by personnel if used regularly and appropriately. When JHA is applied appropriately, the interrelationships between job steps and tasks are better understood, and organizational work dynamics improves.[4]

7.3 JHA INFORMED EXPOSURE ASSESSMENT

The information gleaned from the JHA can be used to form the basis of conducting a full exposure assessment to analyze and quantify potential exposure for chemicals, physical or biological hazards. Typically, a JHA is completed having the input of a diverse team of knowledgeable workers. This collage of knowledge can be key in exposing potential hazards and exposures that may not be identified if the evaluation is being completed by an individual or a small group of workers having minimal knowledge of the process or task. For example, during the performance of the JHA it was revealed that a potential hazard of concern that may occur is the release of an organic substance during removal of a pipe and jetting of a sump pump. An exposure assessment of the task to determine potential exposure to workers that includes development of a sampling strategy, collecting samples, analyzing results, and documenting potential exposure is warranted. The benefits of a comprehensive JHA cannot be understated. A comprehensive JHA can be instrumental in providing potential hazards in a system or work task, and the exposure assessment process provides the mechanism to evaluate and quantify the potential exposure to workers.

7.4 PROCESS REVIEW AND ANALYSIS

A JHA is one of the excellent techniques to assess and implement suitable work processes. It is one of the components of a wider safety and health management system obligations. By scrutinizing the operations of the workplace, establishing appropriate job procedures, and ensuring that each employee receives appropriate training, JHAs can contribute to the reduction of illnesses and injuries, safer working practices, lower workers' compensation expenditures, and higher worker productivity. Priority must be given to new jobs tasks, modifications in processes and procedures, complex job tasks, and jobs with the highest injury or sickness rates. Existing hazard controls may not be sufficient and require further review. In which case, perform a preliminary job analysis. List as a priority job that provides unacceptable risks based on the likelihood of occurrence and severity of impact. Perform a complete breakdown of the work's phases or tasks. Photographing or videotaping the worker performing the task may be useful in completing a job hazard analysis.[1]

A critical review of the process in which maintenance will be performed is a good avenue for determining potential hazards that may be encountered by workers. Analysis of the process can provide information such as past performance issues and repairs, previous work performed on the system, and the hazards encountered during that repair. During the process analysis each step in the process is examined to understand how it works and areas of improvement and identification of hazards workers may be exposed to during the course of conducting work.

7.5 JOB HAZARD ANALYSIS PROCESS REDESIGN

The JHA process has been expanded, redesigned, and incorporated into other work management processes and systems such as the Work Control Process, the Job Safety and Environmental Analysis (JSEA), and the Safe Work Method Statement (SWMS). Each of these processes is discussed below.

Work Control: Integrating the JHA and Work Planning process is an approach of delegating responsibility for workplace safety to those in charge of planning and conducting the task. Hazard controls are included in the Work Planning documentation, and the Safety Group would examine these documents to verify that they are adequate.

This process entails incorporating safety and health into job planning and execution. The goal is to ensure that work is managed and completed in compliance with a documented Safety Management System that:

- Defines the scope of work
- Identifies and analyzes hazards associated with the work
- Develops and implements hazard controls
- Performs work within controls
- Provides feedback on control adequacy and continues to improve safety management

Work control documentation, procedures, JHAs, manuals, policies, and training and qualification records should be examined at some frequency by an assessment team to validate effectiveness of hazard recognition, controls, and mitigation. The assessment team also conducts interviews with key employees responsible for creating and executing the task and the observed work activities and walks through important areas of the work environment. These task-based JHAs are written task-level hazard assessments that form the core of word control documents and contain hazards and controls, boundary conditions, prerequisites, action statements, pre-job talking points, and task notes.[5]

7.5.1 Post-Job Review of JHAs

The purpose of the post-job review is to close out the job and capture lessons learned whether the JHA led to improved work practices or adverse work practices and if instituted controls were adequate. This is achieved through feedback from employees regarding the activity completed.

Post-job reviews of JHAs should be performed with results communicated to employees and supervision. The review should involve employees that completed the work tasks and Safety and Health staff where through discussions and interactions, data is collected in the form of feedback to assess and improve the work process. Feedback from employees should cover any problems, concerns, inadequacies, coordination difficulties, unforeseen events, and/or near-misses that were experienced, as well as how they were resolved. The lessons learned should be incorporated into the task work document for future work.

Job Safety and Environmental Analysis (JSEA): This is a written document that outlines the high-risk work activities to be performed at a workplace, the hazards and risks associated with these activities, and the risk-control measures to be implemented. A JSEA evaluates both environmental and health hazards. The requirements of an environmental analysis and the JHA are combined into one document. Considering the environmental hazards and controls when completing a JSEA or JHA is very important. Environmental hazards are included based on identification and assessment of the work environment. Notable environmental hazards include waste generation, disposal or transport of wastes from a work site, release of air emissions, and spills to water or stormwater.

Safe Work Method Statement (SWMS): The primary goal of an SWMS is to assist everyone on a construction site in understanding what risk controls are in place to safely carry out high-risk construction activities. SWMS establishes a shared understanding among all stakeholders participating in the project or task, and this is documented for easy reference. This document can also be used for training and auditing purposes.

SWMS is used in certain countries to outline the high-risk construction work activities to be performed at a workplace, the hazards associated with these activities, and the risk-control measures to be implemented. An SWMS differs from other documents that focus on specific tasks or processes, such as a JHA or a Safety Operating Procedure. SWMS is not intended to be a procedure but a tool that assists supervisors and employees in confirming and monitoring the control measures necessary in the workplace.

7.6 JOB HAZARD ANALYSIS PROGRAM

The JHA program's goal is to determine the training, equipment, work practices, and monitoring required to safely execute job duties in line with applicable workplace safety requirements. This program applies to all workers who perform tasks that require a JHA. JHA provides findings and recommendations associated with each work task. PPE and other protective measures may be recommended based on the findings. The Safety and Health Professional should assess the JHA Program on a yearly basis. The yearly evaluation will contain all program-related documentation, including completed JHAs. When new activities, processes, and/or roles are introduced, amended, or altered in a way that affects PPE, or the way work is performed, the JHA Program is promptly updated to reflect these changes.

JHA program criteria for businesses that employers should include:

- Reviewing job-specific hazards
- Conducting hazard assessments
- Instituting controls and safe work practices
- Certifying that assessments have been made
- Providing training to all required employees

7.6.1 JHA Data Collection

Collecting information on accidents and near-misses, their impacts and consequences, potential causes, and risk factors is essential to hazard analysis. Data collection can also involve direct observation of job tasks, video and photographic records of job tasks, and accident records. Collection of data can also occur through questionnaires and interviews and through focused employee group sessions. Effective work hazard evaluations need ongoing documentation of these types of data. Ensure that efforts to eliminate the hazards and implement hazard control measures are focused on the key causes of the hazard.[1]

7.6.2 Data Analysis

An effective job safety analysis depends on the ability to perform a good data analysis. Unreliable or poor data input to the JHA makes analysis more challenging and the results potentially unreliable. If hazard control measures that are identified are not integrated into the job task, the information from a job hazard analysis is not useful. JHA has potential problems when it comes to gathering and analysis of task data that can arise from focusing on the sequence of work tasks and steps and the behavioral component of workers with respect to the work environment if data is missed or overlooked.

There are many types of hazard controls, and some controls used to mitigate risks are more effective than others. Engineering controls come before administrative controls and personal protective equipment in terms of effectiveness and efficiency according to the hierarchy of controls. The hierarchy of controls in a Job Hazard Analysis can be achieved by substitution for the hazard, eliminating the hazard, engineering redesigns, or containing the hazard, revising, and modifying work procedures, and reducing job frequency and personal protective equipment.

7.6.3 Data Interpretation

Organizations utilize JHA data developed with input from workers depicting the work environment to design safe work environments. The information gleaned from JHA can also be used to guide worker behavior when conducting work tasks and can be used to improve productivity. When used appropriately the data can also track equipment usage patterns, the effects of varied work schedules, employee safety compliance, and correct PPE use. Long-term data gathering allows organizations to discover various work trends and how they impact workplace safety. Patterns, trends, and anomalies seen through long-term data gathering can disclose possible hazards and risks can be addressed. Such information is critical for raising safety awareness and finding the appropriate preventive measures.

Data analysis is simply a technique of working with data to identify and catalogue useful information that can be used to make decisions based on the data acquired from numerous JHAs. Historical data may be utilized to construct models for anticipating future safety threats based on current work patterns. In addition, the analysis can demonstrate a relationship between job duties and workplace mishaps/ hazards or injuries. Predictive analytics is useful for evaluating risk variables and associated risk factors. It forecasts the likelihood of incidents, aids in the identification of appropriate mitigation measures, and may be used to develop data-driven safety initiatives. Analysis of data provides information that can be used to improve real-time monitoring of associated workers performing work at various locations within a facility.

7.7 PRIORITIZING WORK TASKS

Methods for prioritizing work tasks to determine the order in which a JHA should be performed when there are competing work tasks are shown in Tables 7.1 and 7.2. Consider the degree of the harm in the task severity table if something goes wrong while doing the task. Consider how frequently the worker is exposed to hazards in the hazard probability table. Take the task severity rating and multiply it by the hazard probability rank. Prioritize the tasks with the highest scores.[6]

TABLE 7.1
Severity Classification

	Task Severity Table	
Score	**Classification**	**Description**
4	Catastrophic	May cause death
3	Critical	May cause severe injury or illness
2	Marginal	May cause minor injury or illness
1	Minor	Will not cause injury or illness

TABLE 7.2
Hazard Probability Classification

	Hazard Probability Table	
Score	**Classification**	**Description**
5	Frequent	Very likely to occur frequently
4	Probable	Probably will occur sometime
3	Occasional	May occur infrequently
2	Remote	Unlikely, but possibly
1	Improbable	So unlikely, it is assumed it will not occur

To prevent harm to workers, it is advisable to perform a hazard analysis for each task before scheduling workers to complete an activity.

7.8 HOW TO ESTABLISH JHA PREVENTIVE ACTIONS

The last stage of a JHA is to identify solutions to eliminate or control the identified hazards. Based on the hierarchy of controls, the commonly recognized measures are[7]:

7.8.1 Eliminate the Hazard

Elimination is the most effective measure. These methods should be used to eliminate the hazards:

- Select a different process or procedure
- Change, modify, or improve an existing process or procedure
- Replace or substitute with less hazardous products
- Improve work environment
- Modify or change equipment or tools

7.8.2 Contain the Hazard

If the hazard cannot be eliminated, contact between the hazard and worker may be avoided by using enclosures, machine guards, worker booths, or other similar equipment.

7.8.3 Revise Work Procedures

Consider adjusting potentially high-risk procedures, changing the sequence of stages, or adding extra measures.

7.8.4 Limit the Exposure

Reducing the frequency of exposure to risk is one strategy to limit exposure. It could be necessary to limit the amount of time the worker is in the work area or to employ the proper personal protective equipment. These procedures are the least preferred and effective and are to be used if no other options are available.

7.8.5 PPE Selection

A JHA should provide the information needed to support the selection of the most appropriate PPE for workers. Other controls should be considered before implementing PPE such as elimination, substitution, and engineering and administrative controls.

7.9 JOB HAZARD ANALYSIS SOFTWARE

JHA software is used frequently to expedite the development of a JHA (Table 7.3), and it should simplify the analysis process and give a systematic and consistent way to perform job hazard analysis by utilizing computer algorithms. The software program is instrumental in keeping an accurate record of past, current, and potential

TABLE 7.3
Example JHA

<table>
<tr><th colspan="4">JHA Sample Template</th></tr>
<tr><td colspan="2">Picture of task/equipment (Optional):</td><td>Task:</td><td></td></tr>
<tr><td colspan="2" rowspan="5"></td><td>Name of Department:</td><td></td></tr>
<tr><td>Job Title(s):</td><td></td></tr>
<tr><td>Analyzed by:</td><td></td></tr>
<tr><td>Date:</td><td></td></tr>
<tr></tr>
<tr><td>TASK</td><td>HAZARDS</td><td colspan="2">CONTROLS</td></tr>
<tr><td>1.</td><td>1.</td><td colspan="2">1.</td></tr>
<tr><td>1.</td><td>1.</td><td colspan="2">1.</td></tr>
<tr><td>1</td><td>1.</td><td colspan="2">1.</td></tr>
<tr><td>1.</td><td>1.</td><td colspan="2">1.</td></tr>
<tr><td colspan="4">Required PPE:</td></tr>
<tr><td colspan="4">1.</td></tr>
<tr><td colspan="4">2.</td></tr>
<tr><td colspan="4">3.</td></tr>
<tr><td colspan="4">4.</td></tr>
<tr><td colspan="4">5.</td></tr>
<tr><td colspan="4">Required/Recommended Trainings:</td></tr>
<tr><td colspan="4">1.</td></tr>
<tr><td colspan="4">2.</td></tr>
<tr><td colspan="4">3.</td></tr>
<tr><td colspan="4">4.</td></tr>
<tr><td colspan="4">5.</td></tr>
</table>

TABLE 7.4
Common Hazards and Description

Hazard	Hazard Description
Chemical (Toxic)	A chemical that exposes a person by absorption through the skin, inhalation, or through the bloodstream that causes illness, disease, or death. The amount of chemical exposure is critical in determining hazardous effects. Check Material Safety Data Sheets (MSDS), and/or OSHA 1910. 1000 for chemical hazard information.
Chemical (Flammable)	A chemical that, when exposed to a heat ignition source, results in combustion. Typically, the lower a chemical's flash point and boiling point, the more flammable the chemical. Check MSDS for flammability information.
Chemical (Corrosive)	A chemical that, when it comes into contact with skin, metal, or other materials, damages the materials. Acids and bases are examples of corrosives.
Explosion (Over pressurization)	Sudden and violent release of a large amount of gas/energy due to a significant pressure difference such as rupture in a boiler or compressed gas cylinder.
Electrical (Shock/Short Circuit)	Contact with exposed conductors or a device that is incorrectly or inadvertently grounded, such as when a metal ladder comes into contact with power lines. 60Hz alternating current (common house current) is very dangerous because it can stop the heart.
Electrical (Fire)	Use of electrical power that results in overheating or arcing to the point of combustion or ignition of flammables, or electrical component damage.
Electrical (Static/ESD)	The moving or rubbing of wool, nylon, other synthetic fibers, and even flowing liquids can generate static electricity. This creates an excess or deficiency of electrons on the surface of material that discharges (spark) to the ground resulting in the ignition of flammables or damage to electronics or the body's nervous system.
Electrical (Loss of Power)	Safety-critical equipment failure as a result of loss of power.
Damage of tissues due to overexertion (strains and sprains) or repetitive motion.	Ergonomics (Strain)
Ergonomics (Human Error)	A system design, procedure, or equipment that is error provocative. (A switch goes up to turn something off.)
Excavation (Collapse)	Soil collapses in a trench or excavation as a result of improper or inadequate shoring. Soil type is critical in determining the hazard likelihood.
Fall (Slip, Trip)	Conditions that result in falls (impacts) from height or traditional walking surfaces (such as slippery floors, poor housekeeping, uneven walking surfaces, exposed ledges, etc.).
Mechanical/ Vibration (Chaffing/Fatigue)	Vibration that can cause damage to nerve endings, or material fatigue that results in a safety-critical failure. (Examples are abraded slings and ropes, weakened hoses and belts.)
Mechanical Failure	Self-explanatory; typically occurs when devices exceed designed capacity or are inadequately maintained.

(*Continued*)

TABLE 7.4 (CONTINUED)
Common Hazards and Description

Hazard	Hazard Description
Mechanical	Skin, muscle, or body part exposed to crushing, caught-between, cutting, tearing, shearing items or equipment.
Noise	Noise levels (>85 dBA 8 hr. TWA) that result in hearing damage or inability to communicate safety-critical information.
Radiation (Ionizing)	Alpha, Beta, Gamma, neutral particles, and X-rays that cause injury (tissue damage) by ionization of cellular components.
Radiation (Non-ionizing)	Ultraviolet, visible light, infrared, and microwaves that cause injury to tissue by thermal or photochemical means.
Struck By (Mass Acceleration)	Accelerated mass that strikes the body causing injury or death. (Examples are falling objects and projectiles.)
Struck Against	Injury to a body part as a result of coming into contact with a surface in which action was initiated by the person. (An example is when a screwdriver slips.)
Temperature Extreme (Heat/Cold)	Temperatures that result in heat stress, exhaustion, or metabolic slowdown such as hypothermia.
Visibility	Lack of lighting or obstructed vision that results in an error or other hazard.
Weather Phenomena (Snow/Rain/Wind/Ice)	Self-explanatory.

risks, as well as remedies applied to each hazard. JHA software provides JHA templates developed in accordance with best practices and industry standards. It is the most efficient and successful method for creating, documenting, and managing JHA worksheets. The software tracks health and safety hazards associated with specific jobs and tasks in a consistent, systematic, and effective manner. The software can be accessed from any device such as a computer, tablet, or phone. Data can be collected and analyzed to show leading and lagging indicators in safety and health trends, and compliance with safety and health regulations. Common hazards that are expected to be captured in are JHA listed in Table 7.4. The hazards listed are not all inclusive, however, shown to provide an example of the variation of hazards and potential exposure channels and impact on workers when exposed

REFERENCES

1. Occupational Safety and Health Administration. Job hazard analysis OSHA 3071, 2002. Retrieved on May 28, 2023.
2. Ghasemi, F., Doosti-Irani, A., and Aghaei, H. Applications, shortcomings, and new advances of Job Safety Analysis (JSA): Findings from a systematic review. Safety and Health at Work, 2023.
3. University of California, Santa Barbara. Job Hazard Analysis, n.d. Retrieved May 28, 2023, from https://www.ehs.ucsb.edu/programs-services/industrial-safety/job-hazard-analysis

4. Roughton, J., and Crutchfield, N. *Job Hazard Analysis: A Guide for Voluntary Compliance and Beyond*. Butterworth-Heinemann, 2011.
5. U.S. Department of Energy. Work Planning and Control Assessment at the Lawrence Livermore National Laboratory, 2020. Retrieved from July 1, 2023, from Work Planning and Control Assessment at the Lawrence Livermore National Laboratory, August 2020 (energy.gov)
6. Washington State Department of Labor & Industries. Job hazard Analysis, 2013. Retrieved June 8, 2023, from https://view.officeapps.live.com/op/view.aspx?src=https%3A%2F%2Fwisha-training.lni.wa.gov%2Ftraining%2Fpresentations%2FJHA.pps&wdOrigin=BROWSELINK
7. Canadian Center for Occupational Health and Safety. Job hazard Analysis, n.d. Retrieved June 10, 2023, from https://www.ccohs.ca/oshanswers/hsprograms/job-haz.html

8 Non-traditional Employee Safety and Health

8.1 NON-TRADITIONAL EMPLOYEES SAFETY AND HEALTH

There is no clear definition of what constitutes non-traditional work; however, there are several definitions that have tried to explain the concept of non-traditional work. Non-traditional workers may also be referred to as contingent workers in some industries or environment. Non-traditional workers traditionally change jobs more than once a year and are not usually employees of one company. They are typically known as freelancers, self-employed, or subcontractors and do not receive benefits from their employers. Non-traditional work arrangements also include gig work, contract work, telecommuting, and working for multiple employers.[1] Individuals may embark on non-traditional work to make extra money or gain experience in different areas.

A non-traditional employee is someone that employers would normally overlook or reject when it comes to filling a position. Some common examples of non-traditional employees include:

- Employees that do not meet desirable occupational standards such as a corresponding degree, direct experience, or job-specific qualifications.
- Employees have no work experience, gaps in their work history, or criminal backgrounds.

Non-traditional employees are sometimes known as "hidden workers" or "skilled through alternative routes" (STAR), and they may have the necessary abilities or potential to execute the job but no method to demonstrate it. Individuals may embark on non-traditional work to make extra money or gain experience in different areas. They may lack job security, their incomes may be volatile, and they generally do not enjoy employer-provided benefits such as health insurance, paid holidays, or retirement plans.

8.1.1 Non-traditional Employment Advantages and Disadvantages

Non-traditional employment has significant impacts on workers and employers that include:

Flexibility: Allows workers to balance work with personal or family responsibilities.

DOI: 10.1201/9781032624389-11

Cost Savings: Employers can save money by decreasing the requirement for full-time employees and their accompanying benefits.

Access to Talent: Increases access to a varied talent pool, including individuals who choose or require flexible work arrangements.

Although non-traditional employment does have many advantages, it also presents some challenges that include:

Job Security: Non-traditional workers may have less job security and fewer benefits than traditional employees.

Regulatory Compliance: It is more difficult to ensure compliance with labor laws and regulations related to non-traditional employment arrangements.

Management Oversight: It is more difficult to effectively manage and integrate non-traditional workers into the organization.

8.1.2 Employment Market Evolution

The job market is changing. Employees are increasingly looking for more flexible hours, varied contracts, and different salaries. Traditional employment was considered full-time work under a contract with a specific employer for an indefinite duration, but a new pattern emerged in the early 1990s with flexible relationships between employers and employees. This new pattern represents a non-traditional employment pattern. Work is often paid for per day or by assignment. In order to accommodate this new pattern, organizations need to adjust their human resources policies to address specific needs, characteristics, and work motivation patterns of non-traditional workers.[2]

Table 8.1 lists a range of non-traditional workers that have different work environments and work dynamics.

Employers must change their recruiting and hiring practices to recognize such abilities and identify people with the proper potential. The employment market has changed significantly in recent decades, driven by a variety of factors such as technology, economic shocks, and shifting social norms. Some of these factors are:

8.1.2.1 Technology

The rise of artificial intelligence (AI), machine learning, and the internet has changed how businesses operate, leading to a need for more employees who have the skills and knowledge to use modern devices and technology effectively.

TABLE 8.1
Typical Names for Non-traditional Workers

Free Agents	1099 worker
Freelancers	Independent Consultants
e-Lancers	Contingent Professionals
Solo Practitioners	Home-Based Business Operators
Seasonal workers	Temporary staff

8.1.2.2 Economic Shocks

The labor market was significantly impacted by both the COVID-19 pandemic recession of 2020 and the Great Recession of 2007–2009. These events may have necessitated changes in work arrangements.

8.1.2.3 Changing Demographics

The change in demographics is seen in the number of women in the workforce has increased, and the number of workers aged 65 and over has also increased.

8.1.2.4 Remote Work

The number of Americans working from home has nearly doubled since 2000.

8.1.2.5 Industry Shifts

The healthcare and social assistance industry has grown, while manufacturing employment has declined.

Some other trends in the evolving employment market include:

- The average person now holds 12 jobs in their lifetime, up from 5 to 7 jobs two decades ago.
- The World Economic Forum estimates that 85% of the jobs that will exist in 2030 have not been invented yet.
- Humans and machines are expected to share the workload more equally by 2025.
- Humans are expected to retain an advantage in tasks like managing, advising, decision-making, reasoning, communicating, and interacting.

What is considered beneficial to the workforce in today's changing world has changed, particularly in the aftermath of pandemics. Individuals used to be attracted to employments that had access to a fitness center, allowing employees to exercise during work hours without having to go out of their way. However, in today's dispersed workforce, with some working from home and others in the office, the emphasis has shifted. Emphasis has shifted to well-being programs such as mental wellness support, support for child and elder care, work location flexibility, increased parental and caregiver leave, and good work-life balance policies. In a 2021 study performed by Calm for Business, 97% of respondents indicated that businesses should try to enhance their employees' mental health, and 76% said that mental health benefits were important to them when selecting a new employment.[3]

Safety and health professionals have a responsibility for non-traditional employees, including contingent and subcontract employees and gig workers. Workers may hold multiple ongoing or temporary positions with poorly identified workplace exposures in addition to nonoccupational exposures. The AIHA is particularly concerned about temporary and contract employees who are placed in hazardous jobs without receiving adequate safety and health training.

8.2 GIG EMPLOYEES SAFETY AND HEALTH

Prior to the internet and social media, non-traditional, short-term, and contract labor existed, but the gig economy introduced a new means of linking workers with customers and those looking to hire them. Gig workers are workers who find work using digital labor platforms such as Upwork, Uber, Amazon's Mechanical Turk (MTurk), and Taskrabbit. These and other labor platforms differ significantly in terms of target markets and configurations, but they all function as middlemen between consumers or clients and individuals who perform services. Some people can work full-time hours on one of these platforms for years, while others utilize it as a side hustle. Technically, they are considered self-employed, even though the design and practices of many platforms may encourage individuals to believe otherwise. The workers could work when and where they want, whether remotely or in person.[4]

About one in seven gig workers (14%) earned less than the federal minimum wage on an hourly basis.[5] During the Covid pandemic, innovative, digital-first job platforms such as DoorDash and Uber Eats grew in prominence as millions of people stayed at home and relied on internet delivery services. According to recent research, the number of people using these sites has also increased.[6]

Gig work is frequently associated with on-demand in-person consumer services such as taxi hailing, food delivery, home repairs, and running errands. Many of these platforms set pay rates and aggressively monitor worker behavior using algorithmic management techniques, all while claiming that the employees are independent partner providers or even entrepreneurs. Other platforms provide workers with the option to determine their own fees or provide differentiated services. Platform-mediated work can be done locally for tasks that require human contact such as driving or caregiving and online tasks anywhere around the world such as transcriptions or graphic design can be done remotely.

The main constraints of gig organizations are the ease with which employees might leave the platform/organization and the lack of formal supervision. A platform provider, through an app or website, provides a totally virtual service in which platform gig workers, by definition, do not have a formal human supervisor.

8.3 CLASSIFICATION OF GIG WORKERS

The determination of whether a gig worker providing services to an employer is an employee or an independent contractor carries significant consequences for both the gig worker and the employer in terms of job protections, tax obligations, and eligibility for employment-based benefits and protections. Classifying gig workers as independent contractors saves platforms significant personnel expenses but shifts risk to individuals; their legal classification has therefore been contested in many industrialized nations for some years.[4] If gig workers are categorized as independent contractors, they will not be covered by federal or state wages laws, anti-discrimination, health and safety, collective bargaining, or other worker protection legislation, as shown in Table 8.2. They do not receive employer-provided health insurance or retirement benefits, and they are not eligible for paid sick or family leave in states

TABLE 8.2
Legal Protections Comparison

Labor Standard	Employee	Independent Contractor
Minimum wage	√	X
Overtime pay	√	X
Unemployment insurance	√	X
Workers' compensation	√	X
Paid sick days	√	X
Paid family leave	√	X
Health and safety protections	√	X
Right to a union	√	X
Discrimination and sexual harassment protections	√	X

Source: Economic Policy Institute. Authors' analysis of current (as of May 2022) federal and state laws.

where such benefits are mandated by law. Independent contractors are also ineligible for unemployment insurance or workers' compensation if they become injured on the job. This places independent contractors in a significantly more vulnerable position than employees when it comes to fundamental workplace rights and safeguards.

Many state governments are debating the status of ride-hailing drivers, notably California, where voters and courts have spoken out. In New York, ride-hailing businesses sought to deal with labor unions opposed to designating drivers as contractors by providing collective bargaining rights and certain job benefits. In Massachusetts, lawsuits have been filed seeking to designate ride-hailing drivers as workers, while other bills seek to classify them as contractors and extend benefits.[7]

Employees have these protections in places where they are statutorily prescribed. Independent contractors do not have these protections in any jurisdiction.

The occupational risks of gig work are unique to the work being done and are shared with workers doing similar work outside of platforms. These might include occupational health concerns such as an increased risk of road accidents for Lyft drivers or musculoskeletal ailments from repeated work on a computer. They also involve the risk of visiting an unknown home to conduct cleaning or care-giving services. These difficulties are exacerbated in countries that lack adequate occupational health rules and enforcement.[8]

There are some workers who have experienced some unpleasant sides of gig platform jobs from being harassed to feeling unsafe. Table 8.3 categorizes gig workers according to their level of skill.

Gig workers, who are independent contractors and not employees of a business, may face unique health and safety challenges such as:

8.3.1 Safety Training

Gig workers may not receive the same safety training as traditional employees, for example, training on workplace violence.

TABLE 8.3
Gig Workers Categories[9]

	Physical (Local)	Virtual (Global)
Low skill	▪ **Transportation and delivery services** – Uber – Lyft – Deliveroo ▪ **Household and personal services** – Taskrabbit – Helpling	▪ **Virtual assistant and data entry** – Fancyhands ▪ **Microwork** – MTurk – Clickworker
High skill	▪ **Specialized services** – Medicast (MD housecalls) – GlamSquad – TakeLessons	▪ **Creative and technical freelance** – Freelancer.com – Upwork – Labmate

8.3.2 Personal Protective Equipment (PPE)

Gig workers may not have access to the same personal protective equipment as traditional employees. PPE is usually supplied by employers to traditional employees.

8.3.3 Reporting Safety Issues

Traditional workers have procedures for reporting safety issues. Gig workers may not know who to report safety concerns to or may choose not to report them.

8.3.4 Safety Enforcement

Gig workers may work in a variety of environments, including remote locations, delivery routes, and freelance assignments making enforcement of safety regulations difficult.

8.3.5 Access to Employer Resources

Gig workers may have limited access to safety training, health benefits, and safety equipment.

8.3.6 Sedentary Activity

Gig workers in transportation and other occupations that require long periods of sitting may face the dangers of prolonged sedentary behaviors that can lead to musculoskeletal disorders.

8.3.7 Prone to Injuries

Gig workers who lift, carry, move, or push heavy objects without proper training, tools, or safety equipment may be injured or develop long-term disabilities. Technology can help improve health and safety in the gig economy. Safety apps, wearable devices, and real-time monitoring systems can help identify hazards, provide immediate assistance, and improve communication between workers and management.

8.4 TELEWORK EMPLOYEES SAFETY AND HEALTH

Telework is defined as conducting work-related duties outside of a regular office environment. Many people who work remotely do so from home-based workplaces that may include a specific location for a desk, computer, and other items needed to execute the duties assigned. When employees work solely from home, they often connect with their managers and coworkers using email, group messaging, and video chat services such as Zoom and Skype. Colleagues may exchange files, attend meetings, and connect with one another frequently using these platforms and messaging services, communicating just as they would in an office setting. Telework is also known as telecommuting.

There are many similarities between remote working and telecommuting. The primary distinction between the two is that, although remote employees typically operate entirely from home or distant places, telecommuters are sometimes required to spend at least one day per week at an office, and in-person check-ins are common. Most organizations require employees to be in positions suitable for teleworking and to be eligible for telework in accordance with the organization's policy on teleworking. Certain job classifications and associated job responsibilities may not be suitable for teleworking. Telework arrangements usually require supervisor approval and review by the organization's human resources office to ensure company policies are followed and the decisions regarding telework are administered equitably among the workforces.

Table 8.4 provides a comparison of work activities for telework and remote work and the applicability to employees' requirements.

There are some safety and health protocols that should be considered to be put in place for telework. Employers face the challenge of ensuring that the work location for teleworkers is reasonably safe. It is not feasible for an employer to visit and inspect the home of a worker to ensure certain safety measures are in place. However, a reasonable approach may be to provide the worker with a list of things to consider to ensure safe work conditions when working away from the office location. Some items for consideration for inclusion on a checklist include the following:

- Ensure that fire protection safety plans are in place.
- Is there a smoke alarm, carbon monoxide detector, and fire extinguisher at the telework location?

TABLE 8.4
Comparison of Telework and Remote Work

Telework	Remote Work
Employees should live in same area or time zone with employer	Employees can live anywhere regardless of time zone
Employees are expected to attend in-person meetings or events	Employees usually attend meetings virtually
Communication is synchronous meaning employees must be online during business hours	Communication is asynchronous meaning employees can work online during business hours in their time zone
Employees build relationships with their colleagues and live in the same location as their employers.	Employees work independently away from their colleagues and anywhere they want

- Does the telework location meet the safety requirements of building and fire codes for work locations that are in a commercial workspace?
- Ensure that the workspace is ergonomically designed.

Telework electrical safety requirements to consider that are traditionally enforced in an employer office setting include having surge protection for electrical equipment, grounded electrical outlets, and extension cords in good condition and well positioned to prevent trip hazards. An important health and safety consideration for employers to address when allowing employees to work from home is who will be accountable for health and safety issues, as well as worker's compensation responsibilities if an employee is hurt. To minimize difficulties, the employer and employee should have a formal agreement that clarifies health and safety issues such as: Will employers be able to visit teleworking locations for safety inspections, or will employees be required to complete a safety checklist (see Table 8.5)? What areas of the telework site will be deemed the "workplace"? How will the employee notify their supervisor of any incident or injury? What methods will be used to investigate incidents? In addressing the company policies and procedures regarding teleworking it is critical the Human Resource and Legal Departments be engaged in setting and implementing all associated policies and procedures. Table 8.6 addresses some telework arrangements. See Table 8.7 for the checklist for telework agreements.

Employees often are required to sign a formal agreement with their supervisor and finish a telework training session offered by their organization before they are allowed to work remotely. There are different types of telework arrangements as shown in Table 8.6.

Table 8.7 contains questions that can be considered when formulating teleworking agreements.

TABLE 8.5
Differences between Employee and Independent Contractors[10]

	Employees	Independent Contractors
Employment status	Works for a company or client	Can work for many clients or companies.
Control	Company has direct and effective control of how and when work is done.	The company has no or only limited control over how and when the work is done.
Tools and equipment	Tools and equipment are usually provided by the payer, who is also responsible for repair, maintenance, and insurance costs.	Independent contractors usually own the tools and equipment necessary to accomplish the work.
Subcontractors and assistants	Employees have to personally do the work they have been assigned and cannot decide whether to hire helpers or assistants without the express consent of the payer.	Independent contractors and freelancers do not personally have to carry out the work for which they have been hired. They can hire another party to complete all or part of the work required.
Financial risk	Employees will generally be reimbursed for any expense incurred in completing their work.	Independent contractors generally take on a measure of financial risk and can incur losses. They often have fixed operating costs relating to operating a workspace or hiring helpers or assistants.
Responsibility for investment and management	An employee usually does not need to make any investment in order to provide the service required by the payer.	Independent contractors may be required to make significant investments to provide the service required.
Opportunity for profit	Even though their pay can vary depending on the terms of their contracts, employees normally do not have the chance to profit from their work.	Independent contractors may have the chance to profit or incur losses from their work. They can set their own prices and normally incur expenses to complete the work.
Benefits	Employees are entitled to benefit plans such as registered pension plans, group accident, health, and dental insurance plans.	Independent contractors are not entitled to benefit plans.

Teleworking can be safe and healthy if employees and employers take steps to ensure a safe and healthy work environment. Many employers mandate their teleworking employees to have certain safety and health requirements such as:

TABLE 8.6
Types of Telework Arrangement[11]

Telework Types	Description
Recurrent telework	Telework which occurs on a regular, recurring basis at an alternate worksite
Intermittent telework	Employees work regularly from their primary work site but would telework for limited periods of time based on specific circumstances or job responsibilities that could be accommodated by teleworking
Situational telework	Not a part of a regular telework schedule and requested based on a specific situation

TABLE 8.7
Checklist for Teleworking Agreements[12]

Is there a statement which outlines what types of jobs or tasks may be done at home, or will each application be handled on a case-by-case basis?

Is there a statement that indicates that except where outlined in the policy that the employees' regular conditions of employment remain the same?

Have arrangements been made for normal supervision of the employee to continue (including frequency of contact between employee and supervisor, ongoing feedback, performance appraisals, and career development)?

Has the exact location (full street address) and contact details for where that employee will be working from been recorded in case of an emergency?

Does having a remote or home office violate municipal zoning regulations?

Is there a statement that the employee will follow all of the company health and safety policies when establishing and working in a remote or home office area?

Will business meetings be held at the employee's residence, a third location, or at the defined work location? If meetings may be held in the home office, a statement that the employee will keep the home office area safe for themselves and others who may enter it may be necessary.

Is there a policy detailing how communication between the employee, co-workers, and customers will function?

Is there an agreement about travel time and mileage allowance, if any, between the employees' home and the defined work location? What arrangements are needed if the employee is sent away from the home location to another work business (but not to report to the defined work location)?

Is there agreement on how the workload will be assigned? How will overtime be approved?

Is there a need for company (proprietary) information to be stored in a locked room, desk, or file cabinet?

Ergonomics: Employees set up their work area with an ergonomic keyboard, mouse, and monitor. Position their monitor to avoid neck strain and use a chair that supports good posture.

Lighting: Employees have enough lighting for reading and working.

Ventilation: Employees have good ventilation. Poor ventilation can cause headaches, lung infections, and sinus problems.

Noise: Employees are encouraged to work in a noise-free environment. A work environment that is too loud can be distracting and may reduce productivity.

Sitting: Employees are encouraged to take time to stand, stretch, and take frequent breaks. Sitting for long periods can lead to health problems like obesity, cardiovascular disease, and muscle weakness.

Communication: Employees are encouraged to create a communication structure with regular check-ins and catch-ups.

Safety: Employees are encouraged to ensure their home has a working smoke detector and fire extinguisher. Have an exit plan in case of an emergency.

8.5 INDEPENDENT CONTRACT EMPLOYEES SAFETY AND HEALTH

An independent contractor is a self-employed individual or business hired to conduct work for or offer services to another entity as a non-employee. The independent contractor is also known as a consultant, free agent, or freelancer. As a result, independent contractors must pay their own Social Security and Medicare taxes. An employer who conducts business with an independent contractor is not required to provide employment benefits such as health insurance and company-sponsored retirement plans, which they typically offer to their employees. In an employer-employee relationship, irrespective of what name is used to define the relationship, the worker is not an independent contractor, and the worker's earnings are typically not subject to self-employment tax. In the state of California, several factors are used to determine whether a worker is an independent contractor or an employee. See Table 8.8 for a checklist that can be used to classify independent contractors. Professionals such as physicians, attorneys, plumbers, and others who provide independent services can be examples of independent contractors.

There are different regulations used by different countries to classify a worker as a contractor or an employee. Contractor misclassification happens, for example, when a company misclassifies a worker as a contractor despite the fact that local regulations require a worker to be categorized as an employee. These regulations can vary from country to country.

In the United States, if an independent contractor earns more than $600 from a single payer, the payer must provide the contractor with a 1099 form outlining their earnings for the year.[13]

In transactions involving an independent contractor, the entity benefiting from their work may direct the outcome of the work but does not engage in how the independent contractor does the actual work. Independent contractors have the distinct advantage of being their own employer. They set the time, place, and method of work while ensuring that quality work is completed within the time frame stipulated. Table 8.8 outlines the differences between independent contractors and regular employees.

Independent contractors are not compensated for work-related injuries. Clients may treat them unfairly or discriminate against them. Independent contractors are

TABLE 8.8
California Checklist for Classifying Independent Contractors

Whether the worker performing services holds themselves out as being engaged in an occupation or business distinct from that of the employer;
Whether the work is a regular or integral part of the employer's business;
Whether the employer or the worker supplies the instrumentalities, tools, and the place for the worker doing the work;
Whether the worker has invested in the business, such as in the equipment or materials required by their task;
Whether the service provided requires a special skill;
The kind of occupation, and whether the work is usually done under the direction of the employer or by a specialist without supervision;
The worker's opportunity for profit or loss depending on their managerial skill;
The length of time for which the services are to be performed;
The degree of permanence of the working relationship;
The method of payment, whether by time or by the job;
Whether the workers hire their own employees;
Whether the employer has a right to fire at will or whether a termination gives rise to an action for breach of contract; and
Whether or not the worker and the potential employer believe they are creating an employer-employee relationship (this may be relevant, but the legal determination of employment status is not based on whether the parties believe they have an employer-employee relationship).

not typically covered by OSHA regulations because they are considered to be self-employed. However, in some work situations, some exceptions exist such as:

8.5.1 Working for a Company that Follows OSHA

If an independent contractor works for an organization that is required to follow OSHA regulations, they are required to have a contractor site safety program.

8.5.2 General Contractor Requirements

A general contractor may require an independent contractor to follow OSHA standards through a contract.

Independent contractors are often required to have:

- **Written Contracts:** Establish written contracts that clearly state the independent contractor is not an employee.
- **Worker's Compensation Insurance:** Require the independent contractor to have and provide proof of current Worker's Compensation Insurance.
- **Schedule:** The company should not set the independent contractor's schedule.

- **Work Practices:** The company should not direct how the work is completed.
- **Tools:** The independent contractor should bring their own tools.

REFERENCES

1. AIHA. Serving the Changing Workforce Resources, n.d., https://www.aiha.org/get-involved/volunteer-groups/content-portfolio-advisory-group/serving-the-changing-workforce
2. Vaiman, V. (Ed.). *Talent Management of Knowledge Workers: Embracing the Non-Traditional Workforce*. Springer, 2010.
3. Herron, A. 12 Unconventional Workplace Benefits Employees Want You To Offer. WebMD Health Services Blog, 2022, https://www.webmdhealthservices.com/blog/non-traditional-workplace-benefits-for-employees/
4. Kuhn, K. M., and Galloway, T. L. Expanding perspectives on gig work and gig workers. *Journal of Managerial Psychology*, 34(4), 186–191, 2019.
5. Zipperer, B., McNicholas, C., Poydock, M., Schneider, D., and Harknett, K. National survey of gig workers paints a picture of poor working conditions, low pay. *El trimestre económico*, 89(356), 1199–1214, 2022.
6. https://www.cnn.com/2023/07/24/economy/gig-workers-economy-impact-explained/index.html
7. Anderson, M., McClain, C., Faverio, M., and Gelles-Watnick, R. *Americans' Views of Gig Platform Work and Related Policy Issues*, Pew Research Center, 2021.
8. Bajwa, U., Gastaldo, D., Di Ruggiero, E., and Knorr, L. The health of workers in the global gig economy. *Globalization and Health*, 14(1), 1–4, 2018.
9. Jabagi, N., Croteau, A. M., Audebrand, L. K., and Marsan, J. Gig-workers' motivation: Thinking beyond carrots and sticks. *Journal of Managerial Psychology*, 34(4), 192–213, 2019.
10. https://www.bdc.ca/en/articles-tools/employees/recruit/are-your-workers-employees-or-independent-contractors
11. https://www.ccboe.com/staff/myccps/benefits-pay/telework
12. https://www.ccohs.ca/oshanswers/hsprograms/telework.html
13. Internal Revenue Service. Reporting Payments to Independent Contractors, https://www.irs.gov/businesses/small-businesses-self-employed/reporting-payments-to-independent-contractors

9 Emergency Service Workers Health and Safety

9.1 EMERGENCY RESPONDERS

Most emergencies are usually related to fire and medical emergencies. Emergency services such as firefighting is a recognized dangerous occupation in the United States because these workers are exposed to different types of hazards leading to acute and chronic injuries, illnesses, disabilities, and death. Emergency response jobs are demanding and for those who select these jobs as their career choice is rewarding. Emergency responders may face stressors such as seeing human suffering, running the risk of injury to themselves, having a heavy work schedule, having to make life-or-death choices, and being away from family. The hazards include physical, extreme heat, chemicals, smoke, carcinogens, and struck by hazards from building collapse. Emergency medical services responders also face many of the hazards that firefighters encounter in addition to exposure to infectious diseases.

The fire departments are taking proactive steps to improve health and safety. These steps address the 16 firefighter life safety initiatives (FLSIs). The 16 FLSIs were jointly developed by representatives of the major fire service constituencies during a Firefighter Safety Summit in the United States. The key principles include monitoring the system in operation and using the information obtained from validating and certifying a certain safety assurance level. Some of these initiatives include:

Supervision: Supervisors are trained and empowered on safety awareness, develop safety skills, and establish and implement physical and medical surveillance of employees.

Attitude: Employees are trained to improve safety awareness and become proactive.

Accountability: Improve individual and organizational accountability for health and safety throughout the fire service.

Fitness and Wellness: There is an increasing commitment to wellness by many fire departments. These commitments include gyms, employee assistance programs, weight reduction, and smoking cessation programs.

Cultural Change: Define and advocate for a safety-related culture shift within the fire service, including leadership, management, supervision, accountability, and personal responsibility.

Public Education: There is an increased professional knowledge base to improve the capacity to collect and analyze firefighting-related problems.

DOI: 10.1201/9781032624389-12

More funding and support are needed for public education, which is an essential fire and life safety initiative.

Psychological Support: Firefighters with their families have access to counseling and psychological support.

Violent Incident Response: National guidelines for responding to violent incidents should be established and promoted.

Health and Safety Controls: When engineering and administrative controls are not effective or not feasible, first responders may need personal protective equipment (PPE) to protect them during natural disasters, biological hazards, accidental releases, and terrorism. PPE can reduce worker exposure to hazards, but they are the last line of defense after engineering and administrative controls. The PPE that employers provide to workers must be the appropriate PPE for the work task and at no cost to the employee. Employees must be properly trained on how to use the PPE.

9.1.1 History of Emergency Services

In 1966, the National Traffic and Motor Vehicle Safety Act was signed into law. This act created the first mandatory federal safety standards for motor vehicles. This was the first effort to focus on improving deficiencies in emergency medical systems (EMS). Emergency Medical Systems equipment today is much different from when EMS began many years ago.

Between 1960 and 1973, the United States' emergency medical care system underwent a fast transformation due to several medical, historical, and social factors. Researchers, policy advocates, and administrators working in the field of emergency medical services today must recognize these forces and their limitations to modify the system so that it can consistently deliver high-quality acute care to all workers, and patients, enhance overall public health through injury control and disease prevention programs, fully participate in disease surveillance, and be ready to respond to any new needs that may arise in the community.[1]

Presently, the modern emergency medical technician (EMT) profession has expanded to include standard EMTs, paramedics, advanced EMTs, and emergency medical responders, encompassing different levels of skill and specialization. Table 9.1 below reviews the important events in the development of emergency services.[2]

9.1.2 Firefighters and Emergency

A firefighter is a first responder who is trained in firefighting. Their primary responsibility is to control and extinguish fires that endanger life and property and also rescue people from confined, unsafe, or dangerous settings. Firefighters have been responding to accidents and medical emergencies for decades; however, they may also administer electric shock and drugs to restart a heart, insert a breathing tube, or remove an accident victim from a crushed vehicle while simultaneously treating their injuries.

TABLE 9.1
Important Events in the Development of Emergency Services

Date	Event
1960	The International Association of Fire Fighters (IAFF) publishes a firefighter injury and death report.
1969	The first national standardized curriculum for emergency medical technicians was published.
1971	The first certifying exam for EMTs was undertaken by 1,520 individuals.
1973	America Burning, the final report of the National Commission on Fire Prevention and Control is published.
1974	The National Fire Protection Association (NFPA) begins publishing an annual Firefighter Injury Report and Firefighter Fatality Report.
1976	The Public Safety Officers' Benefits Act becomes law.
1977	The first edition of the Standard for Fire Department Safety Officer was issued as NFPA 1501 and established a standard for a new role in the fire service, the safety officer.
1986	NFPA 1403, Standard on Live Fire Training Evolutions, is adopted by NFPA.
1987	NFPA 1500, Standard on Fire Department Occupational Safety and Health Program, is adopted by NFPA.
1987	NFPA 1501, Standard for Fire Department Safety Officer, second edition, was published.
1990	Congress enacted the Ryan White Comprehensive AIDS Resources Emergency (CARE) Act.
1992	Congress created the National Fallen Firefighters Foundation.
1996	The Occupational Safety and Health Administration (OSHA) ruled to implement two-in/two-out procedures with OSHA 1910.134.
1997	The International Association of Fire Chiefs (IAFC) and IAFF join together to develop Fire Service Joint Labor Management Wellness Fitness Initiative program.
1998	The Firefighter Fatality Investigation and Prevention Program by the National Institute for Occupational Safety and Health (NIOSH) begins.
1998	OSHA's Respiratory Protection Standard is applied to firefighters.
2001	NFPA 1710, Standard for the Organization and Deployment of Fire Suppression Operations, Emergency Medical Operations, and Special Operations to the Public by Career Fire Departments, is adopted.
2001	NFPA 1710, Standard for the Organization and Deployment of Fire Suppression Operations, Emergency Medical Operations, and Special Operations to the Volunteer by Career Fire Departments, is adopted.
2004	The Firefighter Life Safety Summit occurs, resulting in the development of the 16 FLSIs.
2005	The nonprofit Firefighter Cancer Support Network (FCSN) was established.
2011	The Firefighter Behavioral Health Alliance was established.
2012	The National Fire Academy develops the Fire Service Safety Culture that protects Firefighters.
2013	The National Fallen Firefighters Foundation develops and releases the Stress First Aid for Firefighters and Emergency Medical Services Personnel course.
2018	Implemented Disaster Recovery Reform Act.

The hazards that firefighters encounter include inhaling smoke, and hazardous chemicals, heat stress, building structural collapse hazards, workplace violence, infectious disease exposure, electric shocks and burns, and vehicle collisions. Several control methods are used to protect these workers from health hazards that include the following.

Engineering Controls – Firefighters use engineering controls to reduce the risk of exposure to hazards, such as:

- Isolating firefighters from the hazard or modifying equipment
- Placing ladders early in an incident
- Putting materials or processes in closed systems, such as booths or enclosed machines
- Using general dilution ventilation or local exhaust ventilation
- Using guards around moving parts of machinery
- Using cabinets to contain flammable liquids and vapors

Administrative Controls – Administrative controls for firefighters consist primarily of changes to policies, procedures, planning, schedules, or training that reduce exposure to hazards.

PPE – Firefighters always use and carry protective and self-rescue equipment to offer protection against the inherent hazards involved in battling fires. To guard against inhaling smoke, hazardous chemicals, and superheated gases, they use self-contained breathing apparatuses (SCBAs), which supply air through a full-face mask. Other equipment includes fire helmets, hoods and face protection, goggles and safety glasses, gloves, and firefighter boots.

9.1.3 United States Fire Administration

The United States Fire Administration (USFA) is dedicated to minimizing firefighter fatalities and injuries by providing a safer operational environment for emergency responders. This is accomplished through best practices and recommendations for enhanced technology, procedures, and behaviors that will better equip fire departments to respond to all incidents and limit the effects of fire. The USFA goal includes building a culture of preparedness in the fire and emergency medical services by improving their capacity to identify, prevent, prepare for, and reduce community risks, as well as increasing organizational resilience, and developing partnerships to enhance these services. The goal also includes preparing to respond and recover from all hazardous incidents and ensuring that the USFA is an innovative, knowledgeable, and responsive organization.

9.1.4 National Fire Protection Association

National Fire Protection Association (NFPA) is a global self-funded nonprofit organization dedicated to preventing death, injury, property damage, and economic loss caused by fire, electrical, and associated dangers. The association disseminates information and expertise through over 300 consensus codes and standards, research, training, education, outreach, and advocacy, as well as collaboration with others who have an interest in advancing the NFPA goal. Among the information disseminated by the NFPA is an injury report.

9.1.4.1 NFPA Injury Report

The NFPA firefighter injury report documents firefighter injuries that occurred in the line of duty, injury type, and collisions. The NFPA Firefighter Injury report is categorized into five categories:

- Responding/returning
- Fireground
- Non-fire emergencies
- Training
- Other on-duty

The NFPA Firefighter Injury report is categorized by nature of injury into:

- Burns (fire or chemical)
- Smoke or gas inhalation
- Other respiratory distress
- Burns and smoke inhalation
- Wounds, cuts, bleeding, bruise
- Dislocation, fracture
- Heart attack or stroke
- Strain, sprain, muscular pain
- Thermal stress (frostbite, heat exhaustion)
- Other

Fireground injuries are known to cause the highest number of injuries per type of duty. The causes of fireground injuries are also reported in the NFPA firefighter injury report.

Fireground injuries are classified into eight causes:

- Struck by object
- Stepped on or made contact with an object
- Extreme weather
- Exposure to fire products
- Exposure to chemicals or radiation
- Fell, slipped, jumped
- Overexertion
- Other

The information reported in the injury report can serve as an informative means to focus on improvement areas that can help prevent injuries and facilitate safe response to fire emergencies.

9.2 EMERGENCY SERVICE WORKER

Emergency service workers include first responders, emergency medical technicians, and paramedics. An emergency responder is required to have a license, certificate, permission, or other official acknowledgment for their skill in emergency response and whose help is needed during an emergency. These workers provide pre-hospital emergency medical care.

Health and Safety: Emergency medical services (EMS) workers' jobs put them at risk for on-the-job accidents and diseases. According to research, EMS employees have a high risk of fatal and nonfatal injuries and illnesses. For example they are at risk for strains and sprains because they frequently must lift equipment and patients. They are also exposed to blood and body tissues.

9.2.1 Basic Emergency Services

Mentioned are the primary emergency services that offer basic emergency help and care. Members of the public can summon them directly as needed. These emergency services are the primary services available to the public. Emergency Medical Services (EMS) provide pre-hospital emergency care provided by emergency medical responders, emergency medical technicians, and paramedics. These providers may be paid workers or volunteers. EMS can be provided by:

- Private ambulance companies
- Fire or police departments
- Public EMS agencies
- Hospitals

Police and law enforcement agencies offer law enforcement, crime suppression and prevention, criminal investigation, and public order maintenance. There are State Police and Highway Patrol Agencies, County Sheriff and Local Police Departments, Federal Law Enforcement Agencies, and other law enforcement agencies in the United States.

Fire departments provide basic services such as fire suppression and protection, fire prevention, special operations and technical rescue response services, emergency medical response services, and hazardous materials mitigation. Emergency medical services provide basic emergency medicine, patient transport, and technical rescue assistance.

9.2.2 Specialized Emergency Services

Specialized emergency services are emergency services that are not considered one of the primary emergency services. Table 9.2 below highlights the different types of emergency services.

TABLE 9.2
Emergency Service Workers

Services	Responsibilities
Border guard	Agency responsible for the security of a country's borders. Known also as border control, border security, border protection
Coast guard	Maritime law enforcement and rescue within national waters using cutters, aircraft, and intelligence to detect, intercept, and disrupt dangerous and illegal activities
Conservation officer	Wildlife protection, wildlife law enforcement, hunting and fishing enforcement
Animal control service	Offers animal assistance, and protects animals from abuse, neglect, and cruelty
Bomb squad	Bomb disposal, hazardous material disposal, weapons and munitions disposal
Incident response team	Specialized incident response by a group of people who prepare for and respond to emergencies, such as work interruptions or natural disasters
Lifeguards	Expert swimmer who provides medical aid and rescue at pools, beaches, shorelines, dive spots
Military forces	Highly organized and armed groups such as army, navy, air force, marine corps, coastal guard, and space force involved in national defense, internal security, emergency assistance
National Guard	Responds to domestic emergencies, combat missions, counterdrug efforts, and reconstruction missions
Paramedics	Highly trained medical professionals who provide emergency care to patients and transportation to the hospital and act as a link between the scene of emergency and the healthcare system
Park rangers	Law enforcement in parks and nature reserves. Responsible for the protection and preservation of parks and protected areas
Poison control center	Advice on poison exposure and help with poison emergencies and give information to prevent poisonings
Public utilities	Safeguarding and maintaining utilities and infrastructure
Public works	Public works encompasses engineering, construction, and related activities carried out by government for the benefit of citizens
Search and rescue	Rescue of missing, trapped, or at-risk victims and the search for and provision of aid to individuals who are in distress or imminent danger
Avalanche rescue	Rescue of victims in avalanches
Cave rescue	Rescue of victims in caves and underground areas
Mountain rescue	Rescue of victims on mountains or in wilderness areas
Urban search and rescue	Rescue of victims in an urban area, and this involves the location, extrication, and initial medical stabilization of victims trapped in an urban area, namely structural collapse due to natural disasters, war, terrorism or accidents, mines, and collapsed trenches
Security forces	Responsible for ensuring the safety of all base weapons, property, and personnel from hostile forces
Security guards	Asset protection, property protection, private enforcement, protect property, enforce rules on the property, and deter criminal activity
Ski patrol	Medical aid and rescue at ski resorts and mountain resorts

TABLE 9.3
Examples of Equipment for Enforcement, Firefighter, and EMS

Enforcement Equipment	Firefighting Equipment	EMS Equipment
Body armor	Helmet	Ambulance
Ballistic helmet & shield	Hood	Automatic external defibrillator
Public order helmet & shield	Turnout gear	Patient examination gloves
Less lethal device	Respiratory protection	Needles
Restraints	Personal alert safety system	Respiratory protection

9.2.3 Emergency Services Regulations

Regulations have been promulgated to protect emergency services workers. One of the main occupational hazards affecting these workers is the inhalation of contaminated air caused by smoke, harmful dust, fumes, mists, vapors, gases, and other air contaminants. The OSHA 29 CFR 1910.134 regulation is aimed at preventing exposure to air contaminants using feasible engineering controls, administrative controls, and the use of personal protective equipment (PPE) including respiratory protection. Employers must provide appropriate respiratory protection to supplement engineering controls, especially when controls are not adequate. Respirators that are usually used by emergency service workers include self-contained breathing apparatus (SCBA), and atmosphere supplying respirators for which the breathing air source is designed for the user to carry.

Emergency services workers often work in environments that are extremely dangerous and are referred to as immediately dangerous to life and health (IDLH). In IDLH situations, controls have to be put in place to prevent hazardous exposures to life-threatening environments.

Equipment used by emergency workers may vary depending on the nature of the task or the hazard involved. Table 9.3 below lists some examples of equipment for different tasks involving different hazards in firefighting, enforcement, and emergency medical services.

9.3 EMERGENCY MEDICAL SERVICES SAFETY AND HEALTH

Disaster response depends heavily on EMS workers. There is an inherent danger of diseases and injuries at work due to the work tasks they perform and in the varying environments in which these tasks are performed. EMS personnel are exposed to several safety hazards, such as:

- Lifting hazards – Lifting patients and equipment
- Biohazard exposure – Treating patients with infectious illnesses
- Chemical exposure – Handling hazardous chemicals
- Workplace violence – Potentially violent patients

Control measures are necessary to be in place to protect workers from such hazards. The identified hazards need to be mitigated or abated. These hazards can be controlled by the following:

- Safe lifting practices
- Use personal protective equipment (PPE)
- Wear slip-resistant footwear
- Use seatbelts in ambulances
- Practice de-escalation techniques for potentially violent scenarios
- Promote safe patient-handling techniques
- Protect workers from blood and body fluid exposures
- Prevent slips, trips, and falls
- Improve motor vehicle safety
- Implement workplace violence prevention programs

9.3.1 Engineering Controls for Emergency Medical Technicians (EMT) and Paramedics

Controls are put in place to provide protection for EMTs and Paramedics while performing their job responsibilities. Some of these controls that have been effectively used are the following.

9.3.1.1 Ambulance Transportation

- Isolate the ambulance EMT driver from the patient compartment and keep pass-through doors and windows tightly shut.
- Use ambulances with separate driver and patient compartments that can offer airflow to each section.
- Ensure that the door and window between compartments is closed before bringing the patient on board.
- During transport, vehicle ventilation in both compartments should be in a non-recirculated mode to maximize air changes that reduce potentially infectious particles in the vehicle.
- Some vehicles are equipped with a supplemental recirculating ventilation unit that passes air through HEPA filters before returning it to the vehicle.

9.3.1.2 Transportation in Law Enforcement Vehicles

When possible, transport patients in vehicles equipped with plastic shields to isolate the driver in the front seat from the passenger compartment, usually the back seat.

9.3.2 Administrative Controls for Emergency Medical Technicians (EMT) and Paramedics

When the need arises to utilize administrative controls for EMTs and Paramedics during performance of their work functions in situations where engineering controls are not feasible, consider controls that are most effective.

9.3.2.1 Ambulance Transportation

- Limit the number of providers in the patient compartment to essential personnel to minimize possible worker exposures.
- Family members and other contacts of patients with aerosol transmissible diseases should not ride in the transport vehicle. Facemasks (respirators) should be used if riding in the transport vehicle.

9.3.2.2 Transportation in Law Enforcement Vehicles

- Limit the number of officers in the vehicle to the minimum number required to safely manage individuals being transported and reduce potential worker exposures.
- Communicate with receiving facilities such as hospitals or jails, about potentially infectious individuals in advance of arrival so that receiving staff can take appropriate precautions.
- Employers should train and retrain workers on how to follow established transportation protocols.

9.3.3 Personal Protective Equipment (PPE)

Emergency responders must use proper PPE during their work activities. PPE equipment may vary depending on workers' job tasks, which may change frequently throughout the work shift. PPE used include:

- Gloves
- Gowns
- Eye and face protection – goggles, face shield
- Face masks – N95 respirator or respirator with better APF should be worn when performing aerosol-generating procedures.

9.4 EMERGENCY RESPONDERS BURNOUT

Emergency responders are subjected to burnout resulting from the challenges they face every day. Some of the reasons for these workers experiencing burnout can be facilitated by the following:

- High-stress environment (dealing with life-or-death situations)
- Witnessing traumatic situations (human suffering)
- Long work hours (rotating shifts)
- Inadequate staffing

Emergency responders do experience feelings of extreme exhaustion and being overwhelmed during their work. Table 9.4 highlights signs of burnout and traumatic stress that can be seen in emergency responders.

TABLE 9.4
Signs of Emergency Responder Burnout and Traumatic Stress

Signs of Burnout

Sadness, depression, or apathy
Easily frustrated
Blaming of others, irritability
Lacking feelings, indifferent
Isolation or disconnection from others
Poor self-care (hygiene)
Tired, exhausted, or overwhelmed
Feeling like:

- A failure
- Nothing you can do will help
- You are not doing your job well
- You need alcohol/other drugs to cope

Signs of Traumatic Stress

Excessively worry or fear about something bad happening
Easily startled, or "on guard" all of the time
Physical signs of stress (e.g., racing heart)
Nightmares or recurrent thoughts about the traumatic situation
The feeling that others' trauma is yours

9.4.1 Emergency Responder Self-Care Techniques

Emergency responders can use a variety of self-care techniques to reduce stress and burnout. These techniques include the following:

- Limit working hours to no longer than 12-hour shifts
- Work in teams and limit the amount of time working alone
- Write in a journal
- Talk to family, friends, supervisors, and teammates about your feelings and experiences
- Practice breathing and relaxation techniques
- Maintain a healthy diet and get adequate sleep and exercise
- Know that it is okay to draw boundaries and say "no"
- Avoid or limit caffeine and the use of alcohol

9.5 SPECIALIZED EMERGENCY SERVICES WORKERS

There are other Emergency Service Workers that are not often thought of or included in the category of emergency workers. These workers often experience the same or similar stress-related issues as traditional EMS workers. Some of these workers specialize in animal control, explosive control, border security, search and rescue, etc.

9.5.1 Animal Control Services

Animal control services are involved in matters related to stray animals; sick, injured, or dead animals including wildlife; aggressive animals; animal bites; animal cruelty or neglect; animals in distress; livestock or pet show permits; and animal-related municipal code violations. These workers offer animal assistance; protect animals from abuse, neglect, and cruelty; and ensure public safety around animals. The potential hazards to which workers are exposed related to animals include animal bites, scratches, and kicks. The proper handling and restraint of an animal is the most effective way to protect both the workers and animals. Occupational safety and health considerations are associated primarily with animal care. The job hazard analysis (JHA) or job safety analysis (JSA) and subsequent risk assessment approach provide a proactive path toward injury and illness prevention associated with animal control and care.

9.5.1.1 Controls

Engineering Controls – Architectural barriers within animal locations that control personnel access and flow patterns

Administrative Controls – Administrative controls include establishing policies (such as rest breaks), mandatory training, management strategies (such as rotating work assignments), monitoring, limiting worker exposure, measuring performance, training and education, and housekeeping. The development and implementation of a training program that is both proactive and responsive to changing program requirements is critical in equipping workers with the knowledge and skills needed to work safely in animal control services. Training should be periodic and documented.

PPE – PPE may include bite-resistant gloves, safety glasses and safety shoes, earplugs or muffs, respirators, eye and mucous membrane protective devices, coveralls, and full-body suits.

Occupational Medicine Evaluation – All injuries caused by animal bites, scratches, and other similar incidents should be taken seriously. Trained occupational medical practitioners should decide whether subsequent treatment of the injury is necessary. Program management and safety staff should record, monitor, and review any injuries involving animals regularly. Predisposing factors and potential training gaps should be identified in this review, which will then encourage program improvement to enhance injury awareness and preventive initiatives.[3]

9.5.1.3 Bomb Squad

These workers have primary responsibility for bomb disposal, hazardous material disposal, weapons, and munitions disposal. The Bomb Squad responds to suspected explosive devices and Weapons of Mass Destruction (WMD) incidents. Bomb Squad utilizes specialized equipment to approach, inspect, disarm, or remove all types of explosive devices. Squad members may be Certified Bomb Technicians or HAZMAT Technicians Level trained. The Bomb Squad team often performs:

- Investigation of suspicious packages
- Render safe operations performed on explosive devices
- Disposal of found explosive materials
- Collection and preservation of evidence at bombing scenes
- Post-blast crime scene investigations

Bomb squads are made up of highly trained technicians who respond to different types of incidents from minor to complex events. They have different capabilities and responsibilities such as:

Type I Bomb Squad: A full-time bomb team that is committed to tackling complex incidents. Note that multiple or simultaneous life-threatening or time-sensitive improvised explosive device disposal (IEDD) occurrences utilizing advanced improvised energetic materials, electronic or remote firing systems, and tactical explosive breaching support might be considered a complex incident.

Type II Bomb Squad: A bomb team that can handle a moderate incident and can operate full-time or part-time. A moderate event might involve electronic or remote firing systems, complex improvised energetic materials, and a life-threatening or time-sensitive incident.

Type III Bomb Squad: A bomb team that can handle a minor incident and can operate full-time or part-time. The team must have basic IED render-safe capabilities.

Control Methods

Engineering Control – These controls include Total Containment Vessel (TCV) using remote stand-off capability such as complete robot systems.

Administrative Control – These controls include training provided to squad members in several areas such as explosive breaching operations, suicide bombing, radioactive dispersal devices, heavy vehicle bomb countermeasures, explosive disposal procedures, homemade explosive recognition, and post-blast investigations.

PPE – PPE used by the bomb team includes helmet, SCBA/APR respirator, gloves, advanced bomb suits, and leather boots.

9.5.1.4 Border Guard

Border Guards provide border control, border security, and border protection. The United States Border Patrol (USBP) is a federal law enforcement agency that secures the borders of the United States. The USBP is part of the U.S. Customs and Border Protection (CBP), which is responsible for border management and control. The main responsibility of the border guards is combating illegal entries and human smuggling across the border. Border protection officers do not require additional safety measures beyond those already used to protect workers from the hazards they encounter when doing their regular duties.

9.5.1.5 Control Methods

Engineering Control – Use physical barriers to separate border protection officers and travelers by installing sections of clear plastic shielding configured with a pass-through opening at the bottom of the shielding as barrier protection between workers and travelers. If workers are screening passengers for fever, use contactless thermometers to prevent workers from having the need to touch sick travelers and to maximize the distance that can be kept between workers and travelers.

Administrative Control – The use of safety protocols. Signage that instructs individuals on safe preventive practices may enhance the effectiveness of engineering control.

PPE – PPE used by border guards includes eye protection, face protection, respirators such as N95, and gloves. Respiratory protection may be upgraded depending on the identified hazards.

9.5.1.6 Coast Guard

The Coast Guard is charged with enforcement of Maritime law and rescue within national waters with ultimate responsibility for reducing the impact of maritime deaths and catastrophic disasters. The Coast Guard is the United States' foremost marine first responder, preventing loss of life and property by looking for and rescuing those in danger. They serve to detect, prevent, and disrupt terrorist attacks, and other criminal acts within the US maritime domain. The Coast Guard can quickly mobilize resources to offer an immediate and dependable response to marine emergencies in collaboration with and assistance of private sector partners and federal, state, and local authorities.

Engineering Control – Controls include ventilation to prevent guards from exposure to inhalation of contaminants such as carbon monoxide and gasoline fumes generated from the vessels. Another control is to have all vessels equipped with visual distress signals.

Control Methods

Administrative Control – Administrative controls include the training of Coast Guards on advanced water survival, emergency medical care, and maintenance of life support equipment.

PPE – PPE includes personal flotation device, helmet with attached waterproof light source and reflective tape, Thermal Protective Aids, Exposure Suits, and portable marine VHF radio.

9.5.1.7 Park Rangers

Park rangers are usually the first to respond to emergencies in the parks they oversee, including medical, fire, and other incidents. Park rangers are trained in several emergency response skills, and perform the following:

- Search and rescue: Park rangers may be involved in search and rescue efforts in remote locations or assist with cliff rescue or SCUBA diving.

- Emergency medical assistance: Park rangers may administer emergency medical care, including using medical oxygen and AEDs.
- Wildland firefighting: Park rangers may be involved in wildland fire prevention and suppression.

Park rangers face a variety of hazards such as eye and face injuries, cuts, punctures, lacerations, foot and hand injuries, chemical exposure, and exposure to excessive temperatures.

Control Methods

Engineering Controls – Engineering controls include closing parks to limit access to hazards and eliminate exposure to them. Building boardwalks or redirecting traffic to provide a safer path for pedestrians to avoid hazards.

Administrative Controls – Safe work procedures must be established to reduce the risk

of exposure to workers and patrons. Training should be provided to address all potential occupational exposures by people knowledgeable of the hazards associated with park environment and activities. Rangers should receive regular training and testing in physical defensive tactics and firearms use.

PPE – Protective Eyewear – Park rangers applying pesticides must wear chemical splash goggles or complete face protection while mixing and spraying liquid pesticides. They must also wear fire-resistant clothing, gloves, safety boots, leggings, and chaps. Safety shoes or boots may have features that include heat-resistant soles, metal insoles, steel shanks, and arch and metatarsal protection. The use of gloves made from wire mesh, leather, or canvas is recommended to provide insulation from burns and cuts. Park rangers also use other equipment including firearms and batons.

9.5.1.8 Search and Rescue Responders

The Search and Rescue team performs an essential service of rescuing missing, trapped, or at-risk victims. The type of terrain determines the complexities and the terrain over which a search is conducted. These include ground search and rescue, including the use of search and rescue dogs, mountain rescue, air-sea rescue over water, combat search and rescue on the battlefield, and urban search and rescue in cities. Search and rescue teams are trained and should have skills in technical rescue, de-watering, navigation, boating safety, firefighting, and medical response. The search and rescue teams are involved in natural disasters such as wildfires, flooding, and earthquakes. Search and rescue teams are often involved in the following activities:

1. Physical search and rescue operations in damaged or collapsed structures, flooded areas, and transportation accident scenes
2. Emergency medical care at disaster sites for entrapped survivors
3. Search and rescue operations in a water environment

4. Hazardous materials assessments in affected areas
5. Structural or hazard assessment of buildings needed for immediate occupancy to support disaster relief operations
6. Repairing and stabilizing damaged buildings such as cribbing and shoring them

Control Methods

Engineering Controls – Can include shoring systems designed by Structures Specialists to stabilize structures and provide safe access to trapped people. Monitoring can be performed to monitor movement in a remaining structure to help determine where people are most likely to have survived. Demolition can be done to open up debris piles for searchers.

Administrative Controls – Monitor for signs of heat or cold stress, such as abnormal vital signs, confusion, excessive sweating, and fatigue. Modify work schedules, rotate workers, and bolster teams with more workers.

PPE – Search and rescue teams must safely navigate dangerous situations created by emergencies and disasters. Teams must wear ANSI-rated protective eyewear with side shields to protect their eyes from a variety of airborne allergens, including liquids, dust, gases, and debris. Moreover, there is a risk of direct impacts from objects like glass, metal, stone, and wood that may fall or fly. Protective eyewear should have shatterproof polycarbonate lenses and scratch-resistant coatings. During these occurrences, search and rescue workers most frequently sustain eye injuries such as lacerations, corneal abrasions, and black eyes. Fall protection equipment, like lifelines, should be used when there are fall hazards. Search and rescue teams should wear high visibility to be readily identifiable by other rescue and support workers.

Appropriate respiratory protection should be used for activities with dust exposure, such as silica and cement dust and chemicals.

9.5.1.9 Avalanche Rescue

Avalanche awareness and rescue readiness are essential for workers who work around winter sports and go outside of areas that are strictly regulated, such as ski resorts. The most physically challenging rescue step is shoveling, which requires the worker to move a significant amount of snow rapidly and effectively to facilitate rescue because time is the enemy of the buried avalanche victim. In most incidents, only half of the buried victims are still alive after 30 minutes, and the odds of survival lessen as time goes on. In some cases, victims do survive for many hours under the snow. The causes of death include suffocation, trauma, hypothermia, and drowning.

The potential hazard for rescuers during avalanche rescue missions comes mainly from self-triggered avalanches; therefore, thorough mission planning and critical risk-benefit analysis are essential for minimizing risk.[4]

Avalanche Rescue Steps:

- Watch for the victim as soon as someone gets caught in an avalanche, don't take your eyes off them
- Make sure the danger has passed before entering the avalanche slope
- Scan the surface of the snow for any sign of the victim
- Use avalanche beacons if available and follow the signal as long as it keeps getting stronger until you can pinpoint the victim's location
- Listen for the person as you may hear their shout for help
- Probe the area where the victim is likely to be
- Mark the victim's location and leave a probe in the snow as a guide
- Shovel strategically and dig as quickly as possible
- Uncover the victim's head first as asphyxiation is the primary cause of death in avalanches, begin CPR, if needed, immediately after uncovering the person's head and chest
- Send someone for help immediately only after the victim's head is uncovered

9.5.1.10 Control Methods

Engineering Controls – Rescue kit to locate and dig out anyone involved in an avalanche. Kits include avalanche transceiver, probe pole, and shovel.

Administrative Controls – Notification of Avalanche Centers for readiness, training of avalanche workers and rescuers, and practice avalanche rescue with team regularly.

PPE – Reflectors to increase visibility, avalanche airbags, and helmets to protect against impacts associated with an avalanche.

REFERENCES

1. Shah, M. N. The formation of the emergency medical services system. *American Journal of Public Health*, 96(3), 414–423, 2006.
2. Angle, J. S. *Occupational Safety and Health in the Emergency Services Includes Navigate Advantage Access*. Jones & Bartlett Learning, 2021.
3. Schmitt, J. M., Wilson, D. E., and Raber, J. M. Occupational safety and health. *Management of Animal Care and Use Programs in Research, Education, and Testing*, 279–318, 2017. https://doi.org/10.1201/9781315152189-14
4. Hohlrieder, M., Thaler, S., Wuertl, W., Voelckel, W., Ulmer, H., Brugger, H., and Mair, P. Rescue missions for totally buried avalanche victims: conclusions from 12 years of experience. *High Altitude Medicine & Biology*, 9(3), 229–233, 2008.

10 Measuring Performance

10.1 HOW IMPORTANT ARE METRICS IN TELLING THE STORY?

A company's performance has a direct impact on worker safety, profitability, compliance, and sustainability. Tracking and trending how well a company performs in key areas is essential and should be an ongoing activity for sustainable performance. An overarching characteristic of high-functioning programs is the ability to continually adjust and make improvements based on performance. Tracking performance and the associated issues and trends through the use of metrics can provide invaluable insight into program aspects that may be experiencing issues or performing well and program elements that are out of alignment with the company mission or established strategy.

To track performance that is meaningful to compliance, safety and health, and performance-based programs, relevant metrics must be identified and measured in a manner that identifies true performance and enables an organization to make focused adjustments to program aspects or processes that do not meet expectations, regulatory compliance obligations or external stakeholder expectations. The actual metrics and performance indicators to be used to measure an organization's safety and health performance will be based on the specifics of each individual organization and associated business venture, work environment, and culture. Identifying the right set of metrics can be challenging and most often is an iterative process that requires refinement over time.

Metrics are an important aspect of telling the story of how a program or organization is performing as such metrics are an integral part of the strategic plan for an organization. Some benefits of metrics include the following:

- Provide an indication of the effectiveness of a program or process
- Facilitate improvements since performance generally increases in the areas that are of focus
- Highlight areas that are in need of improvement
- Assist in focusing resources on important aspects of the business

10.2 METRIC COMPONENTS AND TYPES

Metrics are data that represent an organization's performance, skills, and activities as they relate to performance. Many organizational leaders understand the importance of measurements in assessing and ensuring they are moving in the right direction. Measurements are undertaken for various reasons such as ensuring that the organization is profitable, identifying outliers that can negatively impact the business, or

DOI: 10.1201/9781032624389-13

efficient operations. Although health and safety are an important aspect of an organization's endeavor, organizations often do not measure health and safety performance in the same context as other important business functions.

There are many performance-measuring metrics that are available that can provide insight. Management typically focuses on the use of metrics and indicators that will help them assess performance, effectiveness, and productivity. The methods used to measure performance will determine the type of impact it will have on organizational performance and the behavior of its members. It is difficult to measure safety because it is impossible to assess individual behaviors, attitudes, and commitment for safety and the impact on the work environment and coworkers. In the workplace, management often requests workers to become engaged and take responsibility for their safety and the safety of their coworkers. This level of accountability cannot be measured therefore cannot be included in metrics that can provide knowledge that can be used toward improvement. The purpose and limits of metrics must be properly understood before developing various measuring methods. A performance indicator can be any statistic that is used to predict and understand trends. The information provided by these indicators is very influential and can help shape policy, business investments, and strategic goals of an organization. There are various ways to collect performance measures, and there are different types of safety performance. These measures include:

- Leading or lagging indicators
- Outcome or process-oriented
- Results or activity-based measures
- Downstream factors or
- Upstream factors, and/or
- Qualitative or quantitative metrics

Outcome performance measures are implemented after the performance of the work task or activity occurs. The measure focuses on the outcome. This type of performance metric is seen in injury/illness incident rates or workers' compensation claims data. Process-oriented performance measures, on the other hand, are those measures indicative of actions or activities performed. It measures how much a work task or activity has been performed. It should be linked to the outcome performance metric.[1] Examples of these measures are detailed in Table 10.1. The primary types of metrics that are used in developing safety and health metrics tend to be a combination of leading and lagging.

10.2.1 Leading Indicators

Leading indicators are considered to point toward future events. A leading indicator is a measurable set of data that may help to anticipate safety and health trends and forecast appropriate controls. They define what actions are necessary for an organization to achieve its goals with measurable outcomes.

TABLE 10.1
Performance Measures Examples

Measurement	Example
Leading	Safety audit results
Lagging	Frequencies and severity of injuries
Outcome or process-oriented	Customer satisfaction
Results or activity-based measures	Number of employees receiving safety training
Downstream factors	Incomplete hazard analysis
Upstream factors	Culture (internal and external to work)
Qualitative	Employee engagement
Quantitative	Employee retention rate

A leading indicator helps an organization to evaluate:

- What processes can be employed to achieve a goal to higher levels of success?
- What knowledge and skills can a team improve to achieve the desired outcome better?
- What steps can be taken to speed up product development or service?

Leading indicators are useful for developing a good understanding of performance because they can provide information on anticipated future outcomes.

10.2.2 Lagging Indicators

Lagging indicators are viewed as confirming a pattern that is in progress. Lagging indicators can only be known after the event, but that does not indicate relevance or importance. This indicator can clarify and confirm a pattern that is occurring over time. They are very useful for confirming trends and changes in trends. A lagging indicator can measure current production and performance status. A leading indication is dynamic but difficult to measure, whereas a lagging indicator is easy to measure but difficult to change or modify. Lagging indicators are usually triggered by an event such as an incident or accident. They also measure production outcomes and employee training programs. Lagging indicators work best when combined with leading indicators to determine trends and whether outcomes were met.

10.3 WHY SHOULD AN ORGANIZATION TRACK PERFORMANCE METRICS?

Performance metrics can play an important role in any organization's success. Organizations measuring performance through metrics are key to visualizing how employees are working safely, and whether zero incident targets are being met or are practical based on the way the business is aligned, and work is performed. Performance metrics provide valuable information about the health and safety and overall health of the organization. The data provided by these metrics can identify issues and may be utilized to assist in determining what solutions should be implemented to improve the organization's health and safety policies and procedures. Many organizations link performance metrics to their business strategy to avoid the appearance that collecting metrics is not included in the strategic plan of the organization. It may assist an organization in determining which departments are functioning well and which may need to be improved. Tracking performance metrics is essential since it can assist an organization by giving relevant information on what is influencing quality, growth, and profit. The following are some of the advantages of measuring performance using metrics:

- Strategic Appraisal: Analyzing the efficiency and success rate of an organization's processes can reveal which methods are effective and which should be reconsidered.
- Evaluating Employee Performance: Employee performance metrics evaluate employees' productivity and efficiency in meeting defined goals that contribute to an organization's overall success.
- Quality of Work – Work quality metrics are used to measure the quality of employee performance.
- Quantity of Work – Work quantity metrics are used to measure the quantity of items produced. Quality is more difficult to measure than quantity.
- Efficiency – The efficiency metric is determined by combining the data from the quality and quantity metrics.
- Productivity – Employee productivity is a significant performance metric that may impact an organization's operations and assist in accomplishing objectives.
- Safety and Health – measure the frequency of injuries and illness and track conditions that contributed to the injury or illness.

10.4 MEASURING PERFORMANCE

Measuring performance consists of measuring actual performance output against an intended goal and the important attributes that are necessary to achieve the goals. Performance measurement requires the collecting, analyzing, and reporting of data that represent the performance of an organization, a system, a process, practices, or an individual. This requires a top-down approach for setting performance criteria, measuring outcomes, and instituting needed changes. Performance measurement is a vital part of the process of monitoring that should demonstrate how an organization is progressing in various aspects of operations including safety and health. The answers to the following types of questions should be explored before developing performance measures to increase the feasibility for usefulness:

- How is the organization performing with regard to its goals and objectives?
- Does the organization have the resources needed to safely accomplish goals and objectives (people, finances, equipment)?
- Are the needs and expectations of customers being met or being exceeded?
- Are regulatory commitments being met?
- Is safety a core aspect of performing work?
- Is worker training maintained to meet the current and future needs of the organization and to facilitate work safely and compliantly?
- Is there a process in place to support continuous improvement?

Many organizations struggle with identifying, developing, and implementing meaningful performance measures that provide a clear view of exactly how the company is performing. Some issues associated with the development of metrics include the following:

- Too many of them are developed and tracked
- Development of metrics that are backward looking may not account for what will happen in the future
- Little or no forward-looking elements
- The same level of importance given to all metrics
- False sense of security and achievement when they are not measuring what's important
- Metrics are not representative of the program or process in which it is deployed
- They are not reviewed and used in business decisions

Performance measures can drive behaviors that will help achieve results as well as drive behaviors that will derail accurate measurements and reporting because of a focus on the end product. Performance measures should be developed carefully, communicated openly and completely, and utilized in totality to gauge performance. Performance management activities often fail when there is a lack of champions and leadership commitment and a lack of relevancy to the business.

10.5 HOW TO MEASURE SAFETY AND HEALTH PERFORMANCE

Performance in an organization can be evaluated by direct observations of the work performed by employees and their behavior, by gathering information through questionnaires, meetings, completed Job Hazard Analysis, reviews, records, and reports. Information can be gained through the following sources listed in Figure 10.1 to guide the development of health and safety metrics.

1. Assess Current Performance: Organizations can take steps to measure the health and safety perceptions and culture of their employees. This assessment will help identify areas that need improvement.

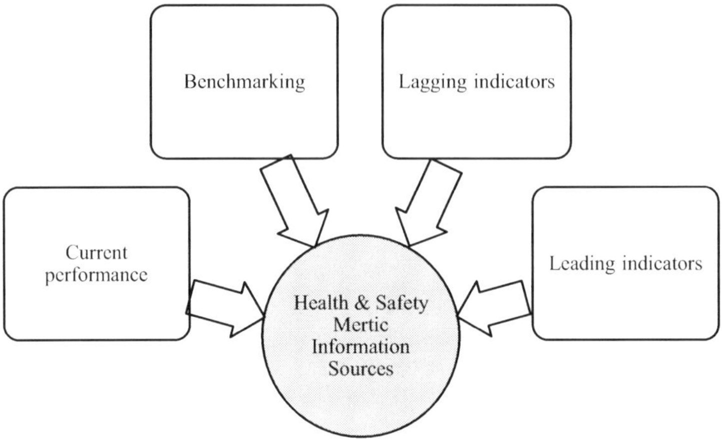

FIGURE 10.1 Health & safety metrics development

2. Benchmark Performance: Organizations can take measurements, and benchmark performance against industry benchmarks or that of competitors.
3. Evaluate Lagging Indicators: measure outcomes after an incident that is effectively a measure of past results. These indicators include incident rate, lost time work injury, etc. Lagging indicators are useful for assessing the present state of workplace safety both in the past and in relation to other businesses. These indicators are calculated based on information acquired on safety incidents and outcomes.
4. Review Lead Indicators: measure activities to prevent or reduce the severity of an incident in the present or future. These indicators include safety training, safety audits, etc. Leading indicators are proactive indicators that give a forecast for the future. They are often referred to as "preventive indicators" because, unlike lagging indicators, they focus on actions that can be introduced or current safety situations that can be changed to prevent future safety occurrences.

When choosing key performance indicators (KPI), it is critical that the goals you specify meet the SMART criteria: the SMART process is pictured in Figure 10.2.

- Specific – goals should be clear and well defined to provide clarity on what the organization is trying to achieve, the importance of the goal, and who is responsible for the KPI
- Measurable – the goal should be measurable using tools that will accurately measure performance
- Achievable – the goal should be achieved using current resources and time to collect and evaluate the data
- Realistic – the goal should align with the safety and health objectives while contributing to the success of an organization
- Timely – the goal should be achieved within a defined time

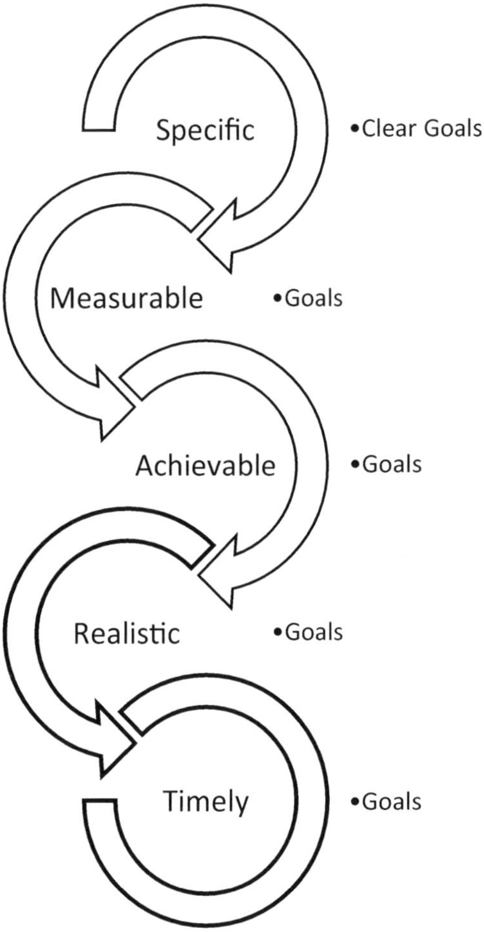

FIGURE 10.2 The SMART cyclic process

There are several factors that must be considered when selecting metrics to evaluate the safety and health program and implementation. The key performance indicator should align with the safety goals and objectives, provide accuracy in results obtained through measurement, and the results must be actionable in providing options for solutions. To determine true progress or improvement, a baseline measurement of the safety and health program should be performed to establish a starting and comparison point.

10.5.1 Examples and Interpretation of Safety and Health Metrics

Developing and implementing health and safety Key Performance Indicators is not only complex, but it may also be difficult to interpret the results and identify the issues, trends, and solutions to reverse trends. A good example of a metric that can be difficult to pinpoint the exact issue and develop the one protective measure to change the outcome into the future is a metric that tracks and trends injuries and illness rates. In this scenario the metrics may indicate that the incidence rates are increasing, and therefore impacting the injuries sustained by workers and the investigation into each incidence revealed a lapse in procedure or process. In this scenario the logical thing to do is to revise the procedure or process to remove the vulnerability. However, if the investigation reveals that the incidence occurred as a result of human performance, it becomes difficult to institute in some cases a solution that will remedy the issue since all employees are not the same. Human performance is difficult to predict and control; however, training and employee engagement in workplace activities go a long way toward closing the gap in risk reduction.

There are a number of metrics used to measure and track the performance of a safety and health program. Some of these metrics are shown in Figures 10.3–10.6 and listed in Table 10.2. KPI reviews the assessments completed and the number of compliance or safety-related issues observed during the assessment. Monitoring safety and noncompliant issues provides an avenue to become aware of potential risks and eliminate issues thereby removing risk factors that can lead to injuries or illness in the work environment. It also provides focus for workers and management on the issues at hand so that they can be cognizant of prevention of future occurrences.

Performance-based incentive (PBI) focuses on the number of first aid cases in conjunction with injuries and illness in the workplace. Keeping appraised of the number of first aid can be important as these cases can be seen as a precursor to more serious workplace injuries. Evaluating first aid cases allows for the opportunity to learn from the incident and enact measures to prevent recurrence and thereby preventing a more serious injury from occurring. Once an employee suffers a severe injury requiring them to seek medical care beyond first aid, the potential for the

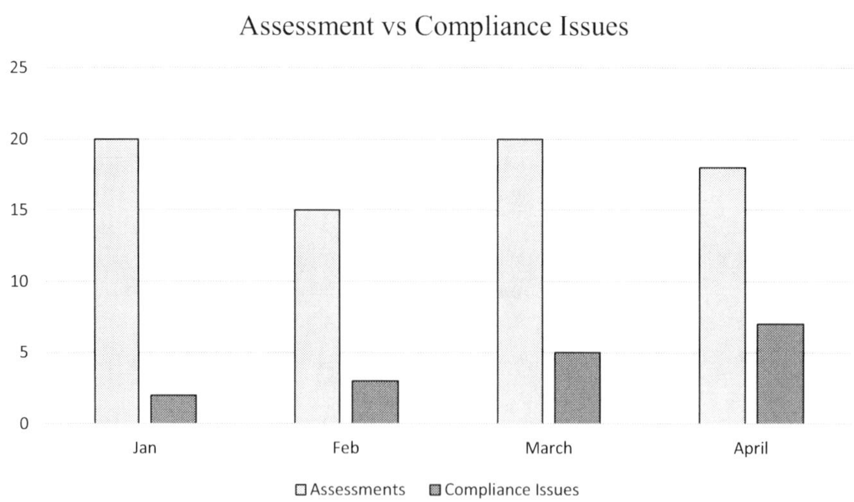

FIGURE 10.3 Assessment vs compliance issues

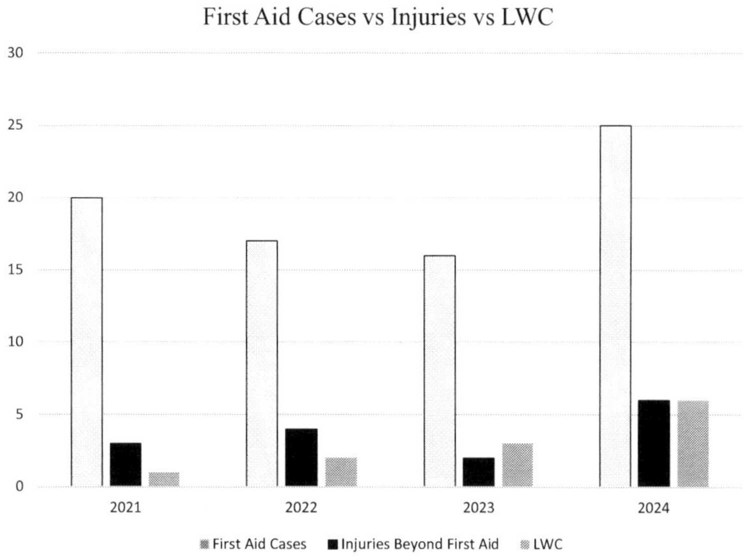

FIGURE 10.4 First aid vs injuries vs LWC

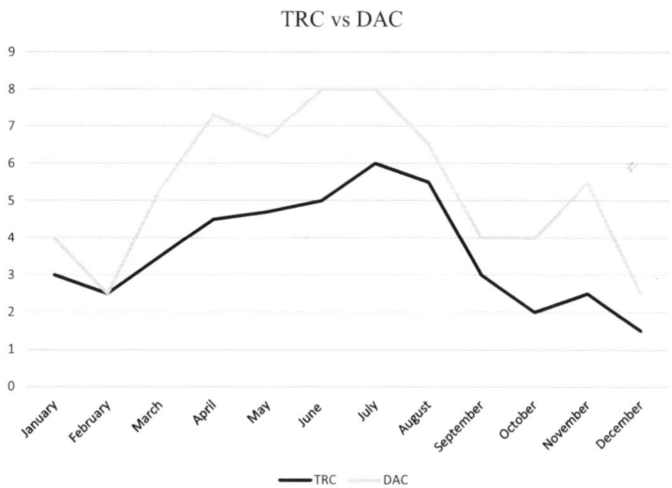

FIGURE 10.5 TRC vs DAC

employee to spend time away from work or restricted from performing their normal work activities while recovering becomes greater.

Figure 10.5 provides a metric that is used to evaluate Total Recordable Cases (TRC), and the potential correlation between those cases that result in days away from work can provide intelligence into measures that can be used to reduce injuries and lower the organization's TRC. A low TRC can provide additional business

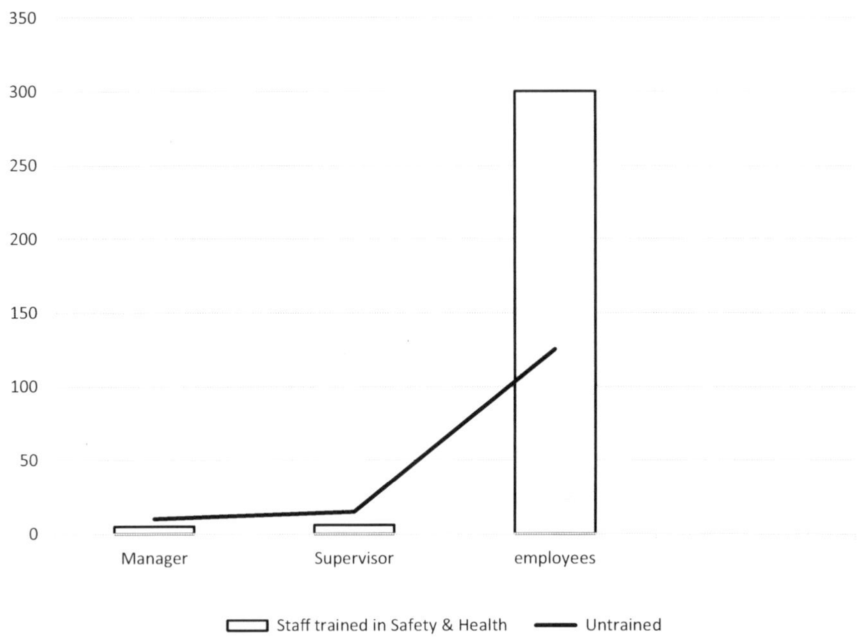

FIGURE 10.6 Safety and health training

TABLE 10.2
Safety and Health PBI

Performance Base Metric	Expected Benefit
The amount of time it takes to resolve issues	Risk management
Employee safety observations performed monthly	Increase employee engagement and accountability for safety
Number of safety violations observed during monthly inspections	Identifying resolving and tracking violations can add the focus necessary to prevent reoccurrence
Safety meeting attendance each month	Increase employee knowledge and awareness of safety performance, procedural, and regulatory changes, and enforce expectation for safe work performance
Safety drills performed annually	Prepare workers on the protective measures to take during emergencies
Safety inspections performed monthly	Identification of issues so that they can be resolved before an injury occurs – risk reduction
Near miss occurring monthly	Identify potential trends, and areas of concern in safety procedures or equipment so that they can be addressed

opportunities for a company. For example, in some marketplaces, when bidding on contracts the contracting organization may request that the company provide their TRC rates for previous projects that demonstrate their commitment to working safety and reducing risk. Therefore, a low TRC can be viewed as a competitive advantage for an organization depending on the business and the market in which they compete. In addition, a low TRC rate can indicate that work is being done safely and worker injuries are minimized. A low TRC is indicative of workers being injured less frequently and potentially losing work time as a result of spending time away recuperating from injuries that occurred while in the work area.

A review of training compliance can provide insight into the importance placed on safety by management and employees. Safety training (Figure 10.6) is an important aspect of improving workplace safety, ensuring compliance with applicable regulatory requirements, and worker proficiency for performing work safely. Tracking and trending training completed for workers and supervisors provide some assurance that the workforce is receiving the information that forms the knowledge needed to perform work safely, improve quality, and improve productivity. It also ensures that regulatory training that is required to be completed at some frequency is completed as required. For example, fire extinguisher training, respiratory protection, bloodborne pathogens, and carcinogens are required annually for impacted employees.

There are many metrics that can be developed to measure safety performance in an organization and if designed to fit the goals of the organization it can provide useful information to improve performance. Other performance metrics that can be considered for usage are listed in Table 10.2 along with some expected benefits.

10.6 DEVELOPING PERFORMANCE METRICS THAT DEMONSTRATES ACTUAL PERFORMANCE

Astute businesses also see performance indicators as roadmaps for their employees. Effective performance measures guide focus and ensure employees are working toward the same goals and purpose. Businesses struggle to choose effective performance measures despite the numerous benefits because of inappropriate design, lack of top-level management commitment, no linkage between metrics and processes or goals, and a poor performance management culture. To develop effective and functional performance metrics, management needs to form a performance metric team[2] and ensure that they have a clear mission, goals, and objectives.

Developing performance measures takes careful thought and consideration for an organization if the process is expected to be effective and serves the purpose in which it is being designed. Too often developed metrics are not effective in providing the information needed to inform decisions on various aspects of the business such as safety and health performance. Therefore, data is collected and placed on the shelf to present to auditors and board members when requested. Table 10.3 lists data collection methods that can be used to inform PBIs.

The most effective measures to use when measuring the performance of a safety and health program can vary for organizations even if they are competing in the same marketplace. Therefore, performance measurement is organization specific, although it can be created based on other effective programs.

TABLE 10.3
Data Collection Methods that Inform Performance Measures (Metrics)

Method	Collection Methods	Examples and Benefits
Survey	Administering a survey with questions to gauge perception from specified individuals. Surveys can be conducted in person, over the phone, or online.	Perception survey data collected to determine employee engagement. The results can be used to increase involvement by employees in various aspects of an organization, for example, hazard recognition and reporting.
Interviews	One-on-one conversation between the interviewer and the interviewee. Interviews can be conducted in person or over the phone.	Conducting an interview of a worker that was injured during work performance. The knowledge gained from a firsthand account can be used to improve process and prevent reoccurrence.
Focus groups	Focus groups are moderated group discussions on specific topics to collect qualitative data on specific topics.	Small group discussion when evaluating culture. Effective when used in evaluating organization culture.
Observations	Involves watching and documenting the behavior of people, as they perform work activities.	Observing a work evolution. Opportunity to see if work is performed by procedure and training.
Secondary data analysis	Using existing data that was collected previously internally from external sources (other organizations) as long as the data is applicable to the scenario being applied to.	Data from another organization concerning the flammability of a chemical. Provides insight into chemical usage and hazards to employees.

Before identifying what is important to measure it is just as important to understand why measurement is necessary. Measurement of performance is a continuous improvement process which involves checking the performance against the standards that demonstrate compliance that was developed to be followed by the entire group or organization. Monitoring performance begins with some questions that may be useful in formulating the metrics needed to gauge performance. Some of these questions are:

1. Why measure?
2. What is important to accomplishing the goals of the business?
3. What is important to customers and stakeholders?
4. What does success look like for an organization?
5. How should and can measuring be performed?

The responses to these questions and others can form the basis or foundation on which to build the performance monitoring program and the type of metrics that should be developed, deployed, and monitored.

10.7 METRICS A FACILITATOR OF CONTINUOUS IMPROVEMENT

Goals of an effective continuous improvement program should include learning from previous activities and events; reduce risk to company and employee; ensure compliance with all applicable regulations and laws; and continue to provide quality products and/or services to customers. A comprehensive improvement program targets all elements of that program at various periodicity. According to Alston, Millikin, and Piispanen,[3] the elements of a continuous improvement program are (1) define a performance baseline, (2) establish improvement goals and objectives, (3) identify and implement actions to meet goal, and (4) measure and evaluate performance. This model is modified and shown below in Figure 10.7 to include the changes that take place during the process at various steps to facilitate continuous improvement. Developing and implementing a Continuous Improvement Plan (CIP) is a good way to guide improvements. A continuous improvement plan should contain a set of targeted activities that are designed to produce gradual changes within a specific area through review, measurement, and analysis of performance data. Without a plan, leaders and workers are not aligned on what to focus on or where improvements are needed or what is most important to the organization. The cycle outlined in Figure 10.1 forms the basis for the continuous improvement plan. Each element should be completely explored throughout the chain, and once the end of the chain has been reached and the information has been communicated to members of the organization, the process is repeated.

When an organization loses the desire to continuously improve the way business is being conducted, they have made the decision to stifle growth, productivity, and

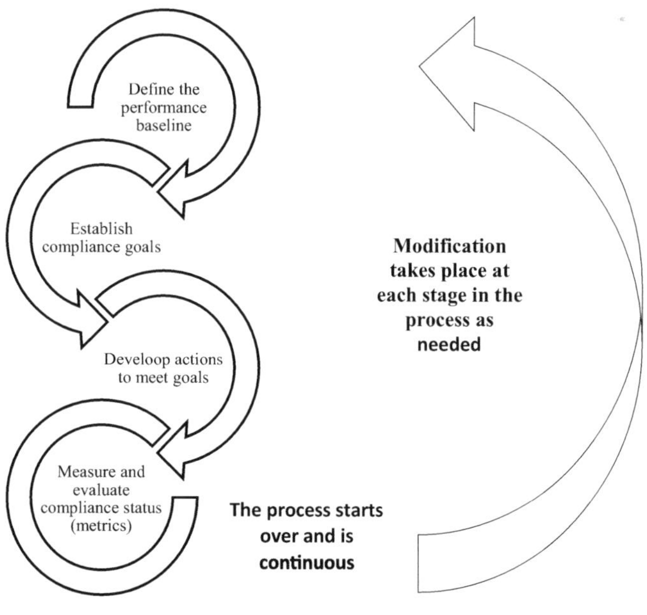

FIGURE 10.7 Continuous improvement cycle

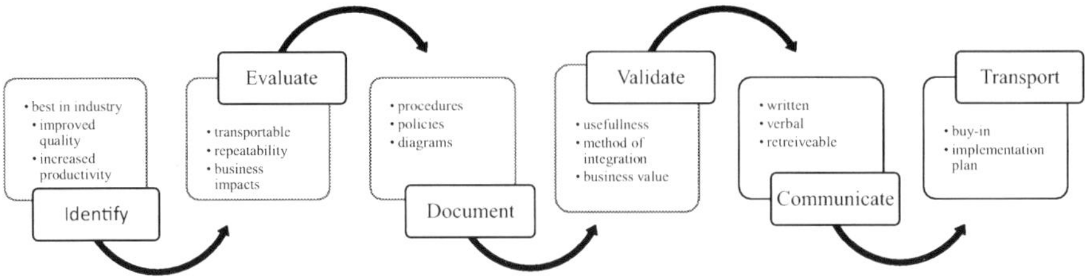

FIGURE 10.8 Lessons learn path to continuous improvement. Note: Transporting what has been discovered to improve your process can be the most difficult step in the process.

sustainability. Continuous improvement should always be at the forefront of all business strategies and workers should be encouraged to participate in the practice of always seeking better and more efficient ways of completing their tasks and improving safety in their environment.

10.7.1 Using Lessons Learned to Enhance Continuous Improvement

Learning from the experiences of the past is just of what lesson learned is about and an important component of continuous improvement. Lessons of the past can facilitate continuous improvement, including lessons that resulted from activities that produced favorable as well as unfavorable conditions and results. Using the knowledge gained from performing activities to avoid the same or similar situation is at the core of the reasoning behind utilizing lessons learned. Effective use of lessons learned involves having a credible process to identify, evaluate, document, validate, and communicate what has been learned and its applicability to the business in which it is being applied. The final step in the process is to transport the best practices to your organization. Figure 10.8 illustrates the path that can be taken to utilize the lessons learned effectively.

In today's business environment, most organizations have a lesson learned program that includes internal lessons learned as well as those learned from other industries or companies that are applicable to their business. In many cases when effectively implemented, this program has assisted companies in averting unpleasant catastrophic events, improving processes, and the quality of services and products provided to their customers.

Each element represented along the path is discussed below.

A. Identify – identify the processes or practices that can be exported and learned by employees that can have positive or negative impacts on the business. The process or procedure slated for transport should be best in industry or can help improve quality and increase productivity.
B. Evaluate – evaluate completely the lessons learned and ensure that it is compatible with the intended use. Ensure that the lessons learned are transportable to the business and can be repeated in a manner to gain optimal results as well as provide positive impact for the business.

C. Document – annotate in detail the lesson learned and the process or procedure that is applicable for implementation. A detailed mapping of the process, program, procedure, or project will be helpful to ensure critical steps are not overlooked and provide ease of understanding.
D. Validate – ensure that the proposed lessons learned are in fact useful to the business or project and can be integrated into the business practices and processes while achieving the desired results.
E. Communicate – communication of lessons learned should be timely and comprehensive through written means; however, verbal communication is acceptable as long as it is followed by written means so that it can be revisited as needed.
F. Transport – ensure that the lesson learned can be transported and used in the manner in which it was observed. The process of successful implementation includes obtaining buy-in from users.

Lesson learned can provide insight that is positive and propelling or it can highlight negative and costly business decisions. Figure 10.9 provides a pictorial view of some potential impacts of positive and negative lessons learned and the potential impact.

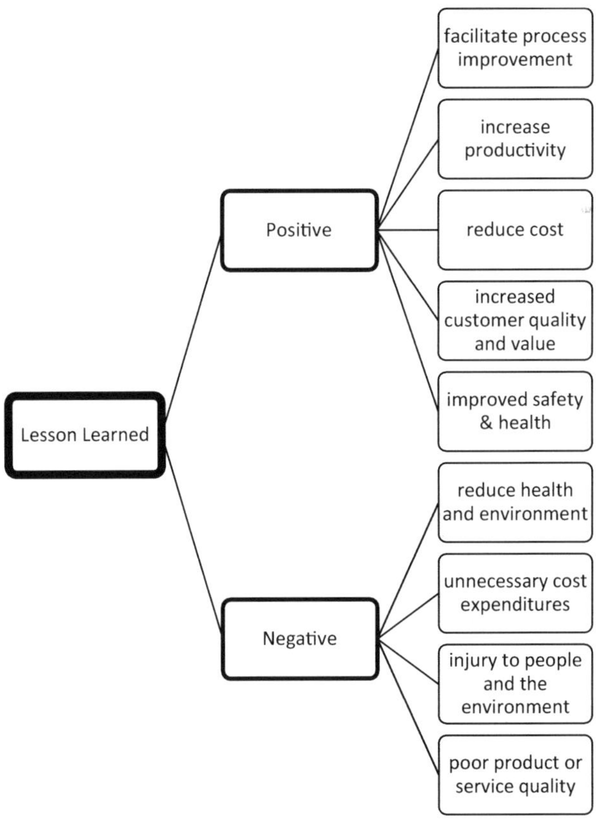

FIGURE 10.9 The lesson learned impact web chart[2]

Positive Lessons Learned: the impact of documenting and publishing positive lessons learned can help a company to replicate the activity or process that can lead to a positive outcome in areas of the company. This can facilitate improving product quality and customer relations, reduce injuries to people and the environment, reduce costs, facilitate process improvement, and increase productivity.

Negative Lessons Learned: this type of lesson learned is typically what is published and shared to assist in not repeating a process or practice that has had negative impact on processes, equipment, people, or the environment. Negative lessons learned often yield impacts to process capability, people, and the environment and are often costly to the organization in terms of downtime for workers and equipment to impact a company's reputation.

Lessons learned provide an opportunity to learn from the mistakes and woes of previous events or from past events encountered by others. A comprehensive lesson learned program is instrumental to the continuous improvement process and can yield best practices for an organization or an industry when modeled and communicated appropriately.

REFERENCES

1. Janicak, C. *Safety Metrics: Tools and Techniques for Measuring Safety Performance*, Government Institutes, 2003.
2. Eckerson, W. W. *Performance Dashboards: Measuring, Monitoring, and Managing Your Business*, John Wiley & Sons, 2010.
3. Frances, Alston, and Millikin, Emily J. *Industrial Hygiene Improving Worker Health Through an Operational Risk Approach*, 2nd Edition, CRC Press Taylor & Francis, 2024.

11 Assessing and Auditing Safety and Health Program Effectiveness

11.1 WHY ASSESS THE SAFETY AND HEALTH PROGRAM?

The question that management is always pondering is what programs and processes are necessary to keep the business afloat and remain profitable for their stakeholders. A part of the decision to do what is necessary to keep the business profitable is to determine how many resources are needed to maintain compliance with regulatory requirements and what regulatory requirements the company is required to ensure compliance with at all times. The answers to these questions ultimately involve investment in resources that include money and people. The primary responsibility of management is to ensure compliance with all applicable laws and regulations to include those pertaining to workforce safety and health. The Occupational Safety and Health Administration (OSHA) places responsibilities on companies to provide a workplace that is reasonably safe for all workers. Once the decision has been made as to the applicable regulatory requirements and the company safety and health program has been established, an initial evaluation should be conducted to establish a baseline and determine if the newly developed and implemented program is appropriate and implemented as intended. This evaluation is often performed in the form of an assessment. Before embarking upon an assessment, care should be given to the development of the goals, strategy, and the assessment approach. Having a strategy that guides the assessment is the best way to ensure that the program is evaluated appropriately, and that the evaluation can determine or demonstrate compliance and is effective in identifying areas that are not in compliance and issues that need to be resolved to include areas of improvement.

Developing and implementing an effective strategy can be beneficial for many reasons including the following:

- Ensure compliance with regulatory requirements
- Document of the health and safety environment of the company
- Can demonstrate that the company values and respect laws and regulations
- Protect the health and safety of their employees
- Protection of the environment
- Protect the company against lawsuits
- Help organizations demonstrate due diligence

DOI: 10.1201/9781032624389-14

- Help build a compliance culture and facilitate building and maintain trust among employees and customers
- Demonstrate to employees that compliance with laws and regulations is important to the company
- Demonstrates to customers and stakeholders that management is willing to provide quality compliant products and services

Some companies include in their organization structure at the highest level (Figure 11.1 example) an organization that focuses on regulatory compliance across the company. These compliance functions range from business, financial, and environmental, and safety and health. In a sense this organization has responsibility for all things regulatory. The compliance organization will often seek subject matter expert assistance from other organizations within the company having specific expertise in the area needed. The reporting line demonstrates the importance of the company operating within the applicable regulatory requirements, signaling to others across the company the importance of following and complying with regulations.

Some companies employ an individual responsible or a group of individuals to track and trend the company's compliance programs. Some companies allocate funding to support a senior level corporate representative to facilitate compliance across the company. This position is often called a compliance officer having responsibilities to guide and facilitate a compliance culture supported by senior management and the appropriate policies and procedures. A compliance officer is typically responsible for creating and communicating policies and procedures that facilitate a company's compliance with state and federal laws; monitor compliance and report issues and trends; investigate compliance issues and coordinate with groups within the company to prevent and eliminate noncompliance. Companies that invest in a compliance officer themselves demonstrate by doing so the importance of a compliance culture to the success of a business and the safety of their employees. In companies where a compliance officer is not on staff, this role may be filled with a group of individuals trained in quality assurance and skilled in conducting assessments. Regardless of which approach is used with staffing the function to ensure that the compliance posture of an organization is retained, either approach can achieve its

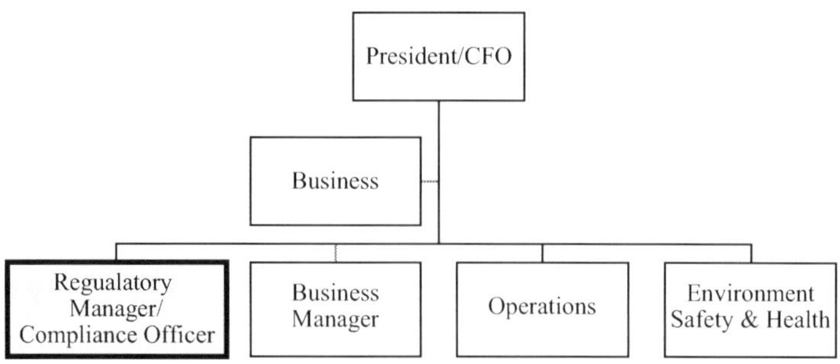

FIGURE 11.1 Example organization structure – regulatory manager

goal of implementing programs that are effective and protect the safety and health of the workforce along with protecting the company from legal exposure.

11.2 EVALUATING THE SAFETY AND HEALTH PROGRAM

There are many ways that an effective evaluation of a safety and health program can be accomplished. The methods selected for conducting an evaluation can be dependent on the type and size of the program, the governing regulation, and the current state of the program, just to name a few elements that can drive the evaluation process. Options that should be considered include conducting the assessment using internal resources, contracting with an external agency with required expertise, or a combination of using internal and external assessors. Regardless of the selected option, a comprehensive assessment strategy should be developed that provides a cross-sectional view of the program's content and functionality. An improperly designed assessment that is implemented can lead to erroneous results that do not accurately depict the status of the program being assessed.

An important aspect of a compliance culture is the role that assessments play in facilitating the safety of the workforce through hazard identification and uncovering programs, policies, and practices that may hinder a safe work culture. The compliance officer, when employed by organizations, has a significant role in facilitating a safe and healthy workplace for employees through ensuring that the programs that are in place are sound and effective in serving the purpose for which they were designed. Evaluating a safety and health program is complex and can be a large-scale project because of the many subprograms that are governed and required by laws and regulations. Table 11.1 shows some of the subprograms that can be included in an overall safety and health program.

The programmatic elements listed in Table 11.1 will not be applicable to all organizations and therefore need not be included in the assessment strategy. Also, not

TABLE 11.1
Safety and Health Program Areas

Asbestos	Confined Space	Carcinogen
Hazardous communication	Beryllium	Lead
Biohazards	COVID	Medical surveillance
Ergonomics	Personal Protective Equipment (PPE)	Respiratory Protection
Walking-working surfaces	Exit routes and emergency planning	Powered platforms, manlifts, and vehicles-mounted work platforms
Ventilation	Occupational noise	Nonionizing radiation
Hazardous materials	Medical and first aid	Fire protection
Machine guarding	Hand and portable power tools	Welding cutting and brazing
Electrical safety	Carcinogens	Vinyl chloride
Benzene	Cadmium	Medical surveillance

TABLE 11.2
Assessment Attributes

Attribute	Focused Results
Compliance	Compliance with applicable regulations
stakeholders	Gauge the perception of stakeholders and their belief of the impact of the program
Financial	Budgeted cost to sustain program
Benefits	Program benefits and overall impact on safety and health of the workforce
Risk	Review of risk profile in its current state and any gap closure activities include impact to risk profile
Quality	The quality of the assessment as well as the quality of recommendations to address gaps
Recommendations	Actions that can improve processes and practices
Schedule	Schedule for performing the program review and schedule for addressing gaps

all elements of a comprehensive safety and health program are listed in Table 11.1. Depending on the company's business strategy and the competing market many more program areas and laws may be applicable. An extensive list and discussion of a solid health and safety program can be found in 29 CFR 1910 and 29 CFR 1926. At a minimum the two regulations should be at the forefront and consulted when determining regulatory applicability and designing a health and safety program. There are many other regulations that may have applicability depending on the business type such as ANSI and ASTM Standards.

After evaluating the program and establishing the current state, it may be necessary to develop an improvement plan to address issues and noncompliance. Section 11.9 provides more detail on developing an improvement plan. It is also prudent to monitor the performance of the program using metrics designed to measure specific elements of interest. Designing metrics has long been a challenge for management because many metrics only lend themselves to measuring what has happened as opposed to what may happen in the future. The goal should be to develop metrics that are forward-looking, although metrics demonstrating a review of past events can also be useful. Developing metrics is not always easy and accurate interpretation of the data is important. A detailed discussion of metrics is found in Chapter 10. Areas for consideration to provide a comprehensive safety and health evaluation should include the following attributes listed in Table 11.2.

11.3 DEVELOPING AND SELECTING AN ASSESSMENT PLATFORM AND STRATEGY

Before embarking upon conducting an assessment, a strategy should be determined and incorporated into a plan that can be implemented as written. Often strategies and plans are developed and written, and implementation is not feasible because of attributes or elements that are not appropriate for the type of assessment or not applicable to the area that is being assessed. When plans are developed and are not implemented a message is immediately sent to the workforce that the issue or program is not important to management or that management has a problem with following through. This perception by workers has far-reaching implications and can extend to other aspects of the business and impact their trust and confidence in the leadership team. Elements to consider when selecting the assessment platform that will be used to form the basis of an assessment plan include but are not limited to:

- The current state of the program
- The program size and critical elements
- Regulatory requirements
- Worker perception of the effectiveness of the program
- Leadership support and involvement

The assessment strategy should be multifaceted to achieve the most effectiveness. What assessment will be performed internally, what assessment will be performed internally will be verified and validated by an external group, and what assessment will be performed by external auditors. Just as important as developing a strategy and conducting an assessment it is important to establish a frequency for reviewing and making adjustments to the plan. Consideration should be given to reviewing and making adjustments to the plan if required at least annually. Minimal components of an assessment strategy are shown in Figure 11.2. Including these attributes at a minimum in a strategy can help guide the focus and purpose of conducting the assessments contained within the strategy.

A brief discussion of each of the components listed in Figure 11.2 is presented in further detail below:

Outcome – Discuss the purpose of the assessment and what is expected to be achieved by conducting the assessment. As a side message not, all outcomes will be positive. However, all outcomes can be beneficial.

Assessment Team – At this point in the process the exact names of the assessment teams may not have been determined. However, what's important is to identify the type of knowledge and experience needed by the assessors

Outcome	Assessment team	Assessment Method	Data collection and analysis	Reporting of results
•Purpose •Goals	•Team member bio •Member area of focus	•Interviews •Document review •Work observance	•Method •Collection •Analysis •Time frame	•Tmeframe to complete report •Presentation of results

FIGURE 11.2 Assessment strategy

to complete assessments. If the assessment team can be identified, then include a bio for each team member.

Assessment Method – Discuss the method that will be used to conduct the assessment. The method can include document reviews, interviews, and/or work observation.

Data Collection and Analysis – Data collection and analysis of the data is directly tied to the assessment method. Data can be collected through a variety of means such as survey distribution, focus group discussion, interviews, documentation review, etc. It may be necessary to take advantage of using quantitative and qualitative methods for analyzing the data.

Reporting of Results – As important as collecting the data is the task of reporting the results in a manner that can be useful in identifying issues and solving problems. Reporting of results should be clear and to the point demonstrating how the area assessed aligned with the assessment criteria and presenting the areas that are in alignment and the areas that are not in alignment.

11.4 IMPLEMENTING AN ASSESSMENT PLAN

Each type of assessment that is listed in the strategy will need to have a written plan detailing specific elements of how the assessment will be completed. There can be significant challenges when attempting to implement an assessment plan. Some of the challenges arise from competing priorities and resources to complete the assessments that are included in the plan. Competing priorities often present themselves in the form of having to complete or address other organizational issues using the same resources that can take the focus of maintaining the plan's schedule. In order to stay on track with completing assessments and other priorities there must be a focus on completing them in a timely manner and as scheduled while avoiding the temptation to reschedule or remove assessments from the schedule. It is incumbent upon management to balance the impacts that competing priorities can have on the assessment plan and schedule.

The other primary challenge that must be balanced is in the form of resources that can also derail the assessment plan and schedule. When planning and implementing the schedule, considerations must be given to the type and availability of the appropriate resources to participate in the assessment teams. Because the schedule is generated often at least a year in advance, it is not always feasible to identify all resources needed and their availability. At plan inception or review the type and availability of resources are often unknown. The actual resource is usually identified and sought when the assessment is closer to needing to be performed. As such, some resources that may be ideal for participation in the assessment may already be prescribed and not have the flexibility to revise schedules that will allow them the ability to participate in performing an assessment.

Some actions that can potentially aid in keeping the assessment plan viable and on schedule include:

Evaluation Area	Assessment Team	Performance Objective	Acceptance Criteria
• Program • Process • Procedure	• Team lead • Assessors	• Lines of Inquiry • Questions	• Regulations • Procedural

FIGURE 11.3 Assessment plan – example

- Keeping the plan visible
- Reviewing the plan monthly
- Planning scheduled assessments at least three months in advance. Planning including scheduling resources
- Including the plan as a part of the company reporting metrics

Minimal components of an assessment plan are shown in Figure 11.3.

Each component of an assessment plan is discussed in more detail below.

Evaluation Area – The evaluation area can be a program, process, or implementation of a procedure. Once the evaluation area has been determined it may be a wise choice not to allow scope creep and maintain the focus and integrity of the assessment.

Assessment Team – Include in the plan the assessment team members' bios and the areas that the team member will be responsible for evaluating.

Performance Objective – The performance objectives are included along with the criteria. These objectives and criteria can be presented in the form of Lines of Inquiries (LOI) or closed-end questions.

Acceptance Criteria – The acceptance criteria should be informed by regulations, policies, or procedural requirements that need to be met to ensure programmatic success.

11.5 INTERNAL PROGRAM ASSESSMENT STRATEGIES

Effective internal assessment programs are the bedrock for leadership to keep abreast of the performance of the company to include the safety and health of the workforce. An assessment program is also a great way to support an environment of continuous improvement through the improvement actions that result from issues uncovered from an assessment. A comprehensive assessment strategy is critical and more importantly, implementation of the strategy is crucial. Often assessment strategies are developed and presented as an important element of how the company operates and shortly after development are placed on the shelf. With the on-the-shelf placement, programs and practices are not assessed or evaluated. In addition, a company may experience breakdowns that will inevitably impact the safety and health of the workforce, the quality of products and or services provided, and even the financial aspects of the business.

11.5.1 Selection of the Assessment Team

When selecting the assessment team care must be provided to ensure that the members have the appropriate skills and credentials to provide valuable input into the actual process and are able to recommend actions that can improve programs and eliminate deficiencies. The assessors should have the applicable knowledge and experience in the topic to be assessed. The assessment team can include all internal members or all external members or a combination of the two. Regardless of the makeup, the selection of team members must be strategic, ensuring that the members selected possess the appropriate skills and knowledge to gain respect and conduct a comprehensive and accurate assessment.

Selection of members from the internal staff to conduct and participate as team members must be done recognizing the importance that the right members be selected. Internal staff are generally known by their peers and may have established relationships or familiarity with selected members that may interfere with assessment performance. Therefore, workers may have formulated thoughts concerning their knowledge and capability to perform a credible assessment and offer solutions to improve performance.

The selection of external members does not have the challenge of familiarity or preconceived notions of individual knowledge or capacity. However, there are challenges that may need to be overcome. For example, workers may believe that external assessors lack knowledge of how work is performed in their industry and work environment. This premise can lead to a lack of respect or desire to participate in responding to questions by the team members and the credibility of the assessment. It is important that the team bios be accessible for those employees who may wonder about the team's capability.

11.6 EXTERNAL ASSESSMENTS TEAM

An external assessment provides a view of a program from the eyes of others who are not directly connected to the business or program being assessed. These assessors typically will possess experience and knowledge of how the same or similar programs are implemented successfully at other companies and can offer alternative ways of accomplishing the same tasks more efficiently. The use of external assessors is a potentially viable means of removing some biases that may be present among internal assessors with respect to the program being evaluated.

Tips to consider when selecting external assessors include the following:

- Review assessment track record and the types of assessments performed
- Request a list of customers where they have performed like assessments
- Discuss with previous customers and gauge their view of the support provided

11.7 GUIDELINES FOR DOCUMENTATION OF ASSESSMENTS

Documentation provides a record of the assessment and contains important information such as:

- What was observed
- The way work was carried out during the assessment
- Measures used to complete work safely
- Employee engagement
- Management engagement
- Procedure compliance
- Regulatory compliance
- Ineffective work practices discovered
- Noteworthy practices

Key components of assessment documentation are shown in Figure 11.4. In addition to what is shown in the figure other aspects may be addressed in the documentation such as noteworthy practices and findings/noncompliance.

A brief discussion of each documentation category is discussed in more detail below:

Were the Acceptance Criteria Met – Discussion and demonstration of whether the acceptance criteria were met or not. If met, no follow-on actions should be required. However, if the criteria were not met, follow-on action should be included in a plan to address and resolve any issues.

Personal Interviews – Summarize the result of the personal interviews and detail any information that was provided to add context to the criteria by which the program was assessed.

Documentation Reviewed – Provide key details on documentation reviewed during the assessment. The documents reviewed can range from procedures, prior assessments, or inspection checklists.

Work Activities Observed – Document the performance of work completed by organization members and their ability to follow work procedures and perform work safely.

Was the acceptance criteria met	Personal interviews	Document reviewed	Work activities observed	Assessment results
• Criteria met • Criteria not met	• Title of interviewes • Number of interviewees by job category	• Procedures • Previous assessment results	• Work planning meetings • Work evolutions	• Findings • Noteworthy practices • Opportunity for improvement

FIGURE 11.4 Assessment documentation

Assessment Results – Assessment results can result in findings which are considered to be actions that are not in compliance with a procedure or regulatory requirement, noteworthy practice, or an opportunity discovered that can improve the efficiency of a process.

11.8 INSPECTION CHECKLISTS

An inspection checklist can be used as a tool to help management determine the areas of the safety and health program that may need to be included in the annual assessment plan as well as used to conduct some variation of an assessment if designed properly. These checklists can be used for safety audits and ensure compliance with regulatory requirements. For example.

- When used for safety audits, it aids in examining and documenting the safety protocols and practices required to ensure that accidents and injuries are identified and mitigated.
- When used to aid in regulatory compliance, it can provide documentation and assurance that regulatory compliance to laws, regulations, and procedures are followed.

Considerations for developing checklists include the following:

- Determine how the checklist will be used and who will be asked to use the checklist
- The checklist should be simple and easily understood
- Ensure that the checklist covers the intended hazards and tasks of interest
- Review examples of checklists that serve similar purpose
- Involve the appropriate subject matter experts in developing and reviewing the checklist
- Can the checklist be automated and is there a process to automate the checklist
- A process for analyzing the data obtained and incorporate issues in the annual assessment plan

11.9 THE ASSESSMENT IN BRIEF (KICK OFF)

The in-brief is an important first meeting because it provides an introduction to the internal SMEs, management, and the assessment team. During the kickoff meeting, it is a good idea for an internal SME or manager to present an overview of the company, process, and/or systems being assessed. This overview provides a first glance into the operations at a high level for the team members. Do not be afraid to present open issues or challenges faced during this presentation. After all, the ultimate goal of the assessment is to gain knowledge in areas that are in need of improvement and the actions that can be taken to improve processes and workforce safety. During the meeting it is a good idea to allow and respond to questions from the assessors,

TABLE 11.3
Kick-Off Meeting

A kick-off meeting is an opportunity to:

- **Set expectations**: Align on goals, deliverables, and how the project will be executed
- **Build rapport**: Introduce the team, and encourage a relaxed and communicative atmosphere
- **Plan for collaboration**: Decide how the team will work together, and how often they'll meet
- **Identify risks**: Discuss potential risks and success factors, and plan mitigation strategies
- **Establish a timeline**: Outline the project's timeline, milestones, and key responsibilities
- **Communicate next steps**: Specify the next steps to build momentum

especially those that are informative and clarifying. Table 11.3 summarizes expectations for the kickoff meeting.

11.10 THE ASSESSMENT OUT BRIEF (CLOSING)

The closing meeting generally is short and to the point. This meeting should be designed to provide an opportunity for the team lead to communicate to the organization's management team and SMEs issues, noteworthy practices, etc. that were noted during the assessment. Although the team lead oversees the meeting and does much of the talking, team members are expected to be present to demonstrate solidarity and, in the event, questions are posed concerning input that they may have intimate knowledge of. Once the out brief concludes the team lead works with the team to develop and deliver the final report. A good practice for report submittal is to provide the written report within a week or two after completing the closing meeting if feasible. Report accuracy and presentation are of utmost importance.

An out-brief should include an executive summary and findings from all areas assessed.

Some recommendations for writing an assessment brief:

- Use clear and concise language
- Provide a detailed description of task requirements
- Specify any format, structure, word count, or guidelines
- Outline expectations about the quality of work

11.11 IMPROVEMENT IMPLEMENTATION PLAN (IIP)

The final step after the assessment has been completed and communicated is the development of a plan to address the issues that were discovered during the assessment and identify and transport any best practices that were identified if any. The IIP can be developed and implemented in various ways depending on the internal systems used for tracking and trending issues or the plan can be as simple as a spreadsheet (see Resource Packet). The plan should list the issues to be addressed, the actions that will be needed to resolve and prevent reoccurrence, the completion

TABLE 11.4
Improvement Plan Elements

Improvement Implementation Plan
An improvement implementation plan can include: • **Objectives**: Define the goals and objectives of the project • **Analysis**: Identify and analyze the existing processes • **Solution**: Develop a plan to address the problem, involving the people responsible for the process • **Evaluation**: Compare the actual outcomes with the expected outcomes, and analyze the differences • **Monitoring and evaluation**: Also known as "M&E," this is a critical part of the program management and implementation cycle • **Resource plan**: Consider resource availability and dependencies to ensure the schedule runs smoothly

date, and the assigned responsible individuals at a minimum. The content of the plan should be shared with organizational members as well as senior management. Just as with implementing an individual improvement plan, the benefits of implementing the improvement plan to address issues discovered during the assessment will provide motivation to workers to perform at their best having knowledge of actions that can set the organization up for better performance and help facilitate a mindset of continuous improvement. Tracking completion of the actions to be performed in support of resolving issues and improving operations should be performed to ensure completion of actions. Elements of an improvement plan are documented in Table 11.4.

Part IV

Chapters 12–15 provide details of some of the program challenges that are faced along with the development and implementation of improvement measures.

DOI: 10.1201/9781032624389-15

12 The Use of Technology to Improve Worker Safety and Health

12.1 WEARABLE TECHNOLOGY

The workplace is evolving and there is an ongoing concern about the shifting workforce and the potential impact of the frequency and severity of workplace injuries. For many organizations, worker safety depends on always knowing the location of their workers and what hazards they are facing, and then using that information to proactively improve safety protocols and procedures.

Employers have an objective to guarantee that employees have a high level of job safety. Because working in an industrial environment is risky, workplace accidents occur daily leading to work injuries. Workers must wear proper personal protective equipment (PPE) to protect themselves from risks where effective safety controls are not adequate or feasible.

Innovation in the wearable technology space has led to enhanced employee safety solutions designed to mitigate work injury risks, increase productivity, improve the overall work experience, and enhance employee retention. Innovations and technology in the workplace have now focused conversations on the status of safety technology being utilized in the workplace, such as wearable devices, artificial intelligence (AI) or computer vision, the Internet of Things (IoT), software applications, and drones.

Smart wearable devices are user-worn accessories with integrated electronics and computer technologies that retrieve or report data. These gadgets, as referred to by some, may gather and store information on the user's physical activity, movement, heart rate, and even speech data. The data collected can be analyzed[1] and used to make decisions on user safety and health risks. The data collected by wearables can assist in health management and be used to improve employee training and ergonomics for workplace wellness and injury prevention programs.

Wearable devices are the future of workplace safety and health, doing much more for workers such as workers tracking their steps around the office or receiving texts during meetings. The current most popular wearable devices are smartwatches, widely favored for their convenience and ability to provide real-time data.

Wearable technology has various advantages, including reduced OSHA recordable, increased productivity, improved employee health management, and lower total workers' compensation claims expenses. Safety wearable devices allow for real-time monitoring of a worker's vital signs and environmental conditions. This functionality

DOI: 10.1201/9781032624389-16

is important as it helps supervisors respond to unusual activity or hazards that are detected, thus preventing accidents before they happen. These wearable devices can greatly increase emergency response times during an emergency, as they offer timely access to critical information that could be lifesaving. A major concern is whether these technologies can assist broad organizational corporate goals while maintaining privacy. Education is important because many wearable technology users unknowingly give away data access, without fully understanding how the information will be used. Transparency about how the wearable device works and how data is maintained builds trust among employees and increases wearable device usage.

To ensure the successful implementation of wearable devices in the workplace, organizations should look for the following features in wearable devices:

- **Easy to Use**: Wearables should be easy to use for workers, practitioners, and OHS professionals. As such there may be a need to create wearables that are tailored to the individual worker's needs and are simple to integrate into their routine.
- **Affordability Is a Key Aspect:** Wearables should be economically priced to promote wider adoption.
- **Compatibility:** Wearable technology should connect easily with workplace processes and routines. Compatibility with workplace processes is critical for practical applications.

Environmental health, and safety (EHS) through wearable technology is going through an evolution in incident detection, avoidance, reporting, and management through the rise of "connected safety" in the workplace.

12.1.1 Wearable Technology Products

Wearable technology products are advanced sensors and computing technologies that a user can wear on their body during daily activity to generate, store, and transmit data. These devices can monitor productivity, identify safety hazards or risks, provide augmented instruction to improve task management, and facilitate health and wellness. The advancement of wearable technology ranges from basic monitors with simple personal alarms to a network of linked devices that may combine worn gas detectors with area monitors and other sensor devices via cloud connectivity. Smart, cloud-based software receives data and uses it for analysis and reporting.

12.1.2 Industrial Wearables Are Usually Divided into Four Groups Based on Function

Monitoring – Wearables may measure a worker's physiological responses, such as body temperature, heart rate, and oxygen levels, as well as ambient factors including air temperature, CO_2 levels, and noise. Wearables may also detect a worker's proximity to other workers as well as possible workplace risks.

Supporting – An exoskeleton is a form of wearable device that supports and assists a worker's mobility or enhances the human body's capabilities. Some exoskeletons may include haptic technology, which may simulate touch by imparting stresses, vibrations, or movements to the user. These may serve to notify the user or enable interactive action.

Training – Wearables capable of providing workers with simulated training include virtual reality (VR) and augmented reality (AR) headgear.

Tracking – Wearables may track the location of a worker.

Table 12.1 below summarizes the different functions of wearable devices.

12.1.3 Positive Safety Culture Supports Wearable Technology

The optimal use adherence and outcomes of wearable technologies largely depend on worker acceptance of the technology. The safety culture of an organization will largely influence the acceptance of wearable technologies by its employees. The optimal culture encourages employee involvement in the discussion of wearable devices and the choosing of devices. Other factors affecting acceptance include performance expectancy (belief that the device will meet its objective or be useful in the workplace), effort expectancy (simple to use), and social influence (perception that management expectation that workers will use the device).

Organizations intending to implement wearable technology should: (a) focus its use on improving and enhancing workplace safety and health, (b) support a positive safety culture, (c) ensure sufficient data, evidence, or documentation to support employees' beliefs that the wearable will meet its intended objective, and (d) engage employees in the process of choosing and implementing wearable technology.[2]

12.1.4 Wearable Technology in Safety and Health

Wearable technologies may be used to improve safety and health in a number of ways. For instance, wearable devices worn by workers can help encourage safe body movements that reduce the risk of overexertion or strains and pulls resulting from awkward posture or lifting preventing risky movements that can lead to health issues such as musculoskeletal disorders.

TABLE 12.1
Wearable Device Functions

Wearable Device Functions
Monitoring devices: Alert workers to changes in vital signs or the workplace environment
Supporting devices: Help workers with tasks like lifting
Training devices: Provide feedback on movements or help improve performance
Tracking devices: Observe the location of employees on the worksite

TABLE 12.2
Wearable Technology in Safety and Health

Wearable Devices

- **Physiological status monitors**
  - Smartwatches and Armbands
  - Monitors a worker's heart rate, breathing rate, and core body temperature
  - Alert workers to signs of stress or overexertion
- **Fall detection devices**:
  - Detect falls and alert supervisors immediately
- **Lumbar motion monitors**:
  - Measure back motion in 3D to identify high-risk motions
- **Exoskeletons**:
  - Help workers hold heavy tools or operate them in awkward positions
- **Sensors**:
  - Detect levels of consciousness and alert the user if they become drowsy
- **ECG tests**:
  - Alert users to health concerns.
- **Step trackers**:
  - Track user's steps

Table 12.2 summarizes the use of wearable technology to enhance safety and health.

AI/Computer Vision – Computer vision and AI systems employ cameras to detect potentially unsafe movements and hazards in the workplace and deliver real-time alerts. An employer can utilize these technologies to review, appraise, and take steps to help avoid worker injuries.

Internet of Things (IoT) – Internet of Things (IoT) sensors are bits of hardware that can monitor and gather data on environmental changes such as temperature, pressure, and motion. When these sensors are connected to a network, they can exchange data. The Industrial Internet of Things (IIoT) refers to a network of interconnected devices for designing, maintaining, monitoring, optimizing, and analyzing industrial operations to gain real-time insights and make more effective decisions. IIoT provides data-driven insights in real time to deliver better outcomes.

The IoT devices range from the simplest wrist-worn activity monitors for tracking worker location and movement to complex robotics-enabled exoskeletons used in construction and warehouse applications. Wearable devices represent a hybrid of humans and technology that can transform and optimize worker safety and efficiency. IoT devices can have different effects on workplace safety since they are divided into different categories and have versions designed for certain tasks.

Software Applications – Software applications can be used with video recordings or other technologies to provide a more comprehensive risk assessment. For example, an employer can utilize a cell phone to record a video of a worker performing various duties, which can then be uploaded to a software program for a detailed analysis of worker actions in real-time situations, enabling employers to identify potential hazards and recommend improvements based on observed behaviors and workplace conditions captured on video recording.

12.2 ADJUSTABLE WORKSTATIONS

Workstations consist of computers precisely designed and configured to meet the worker's technical computing requirements, including high performance, data integrity, reliability, and manageability. Industrial height-adjustable workstations can accommodate a diversified and ever-changing workforce. These workstations have easy height adjustment and encourage the operator to work at an ideal ergonomic height and effortlessly switch between sitting and standing positions. Standing while working is still an uncommon practice among workers, and there is little data to support its benefits.[3] From a health management perspective, increased physical activity and reduced sedentary behavior are important and beneficial.

A height-adjustable workstation allows for adjustment of the work surface height to permit a worker to adjust between sitting and standing during a work activity. Adjustable workstations that are correctly constructed allow workers to sit or stand, as well as modify the keyboard and mouse to match the worker and provide an ideal hand working height. For about 95% of the population, fixed workstations that are constructed at the conventional 29–31-inch height are too high. Workers using workstations will be able to modify the height of the work surface to an optimal working height according to their stature and height with correct table height adjustment, preventing awkward postures that increase the risk of injury. Neck pain, shoulder stiffness, lower back pain, and mental health issues have also been identified as major causes of decreased work performance.

It is not healthy to sit or stand for prolonged periods. Excessive sitting is related to poor work productivity, poor work engagement, and decreased presenteeism and can have an overall negative impact on working conditions. Excessive sitting time by workers has been linked to an increased risk for many adverse health outcomes including mortality.

The worker should adjust posture frequently when using a height-adjustable workstation. Workers are encouraged to sit and stand on a 1:1 ratio such as sitting for 30 minutes, standing for 30 minutes, and repeating the process. The overall recommendation for improving the general health of workers is to reduce sitting hours by four hours per day through standing or participating in low-intensity activities.

Medical research has shown several health hazards are associated with a sedentary lifestyle, including obesity, heart disease, diabetes, and high blood pressure.

Workers spend long hours at work sitting at their desks, putting pressure on their lower back, neck, and shoulders. Studies have shown that alternating between sitting and standing prevents musculoskeletal discomfort and musculoskeletal disorders. On the other hand, workers who add movement like standing during their work duration enjoy better health, wellness, productivity, and satisfaction.

A study published in the *International Journal of Workplace Health Management* on assessing the Health Impacts of Adjustable Workstations found that adjustable workstations that encourage sit-stand activity are associated with increased productivity, better mental concentration, and improved overall health in workers who used adjustable desks over 12 months. Most workers stated that using adjustable workstations improved their health outside of work. About 47% of people with adjustable workstations in this study reported a decrease in discomfort in their neck, shoulders, and upper back. After 12 months, 88% of participants who got adjustable workstations said they were easy to use, 65% said they were more productive, and 65% said they had beneficial effects on their health away from the office. Additionally, those who had adjustable workstations reported better concentration.[4]

It is very important to provide adjustable workstations for employees as this has a lot of benefits for the employer. When employees are comfortable, they are apt to exert their best and become very productive. Adjustable workstations are beneficial to the employees as well as employers. Some of the benefits include the following:

- Reduced musculoskeletal disorders and injuries
- Reduced non-value-added motions
- Reduced workers' compensation costs
- Improved productivity, accuracy, and quality
- Experience virtually maintenance-free operation
- Improved employee satisfaction and morale

12.2.1 Custom Workstations

A custom workstation can be as simple as a small shelf or as elaborate as a multi-station configuration. Custom workstations can be designed to meet the unique needs of an organization, and these needs can be for various purposes such as increasing functionality or productivity. Table 12.3 summarizes when a custom workstation may be considered

Customizable workstations can provide high performance, flexibility, and value. One can choose the components that will give them the best performance for their specific tasks. They may be personalized in many ways, including:

Style: Available choices include a sleek and modern or classic and functional style.

Materials: The selection of materials can be made in collaboration with an expert team.

Layout: Collaboration with an expert team is important to create a customized layout.

TABLE 12.3
Custom Workstation Considerations

Productivity and Accuracy	Custom workstations can help further reduce cycle times and improve accuracy compared to standard modular stations
Safety Culture	Custom workstations are ideal where safety is a top priority
Ergonomic Focus	Custom workstations are recommended for organizations with an ergonomic focus
Employee Well-being	Custom workstations show employees that they are valued in the organization and their well-being is a priority
Lean Initiatives	If the business focuses on LEAN, custom benches can reduce non-value-added motions
Unique Workflow	Custom workstations are recommended where highly specialized tools or material handling are needed to have a seamless integration

Ergonomics: A workstation that supports proper posture and reduces stress on a worker's spine is important.
Modularity: A workstation that is designed for each task is desirable.
Flexibility: A workstation can be customized in many ways, including layout, height, shape, size, and location.

12.2.2 Types of Workstations

Standing Workstations – These workstations are height-adjustable desks that allow employees to switch between sitting and standing positions throughout the day. Standing workstations have a higher surface for the computer monitor and keyboard and may include a footrest. Standing workstations have recently gained some attention due to the perceived potential health benefits of standing for at least part of the workday.

Cluster Workstations – This workstation desk is designed for multiple employees in a cluster to encourage communication and collaboration.

Single Pentagon Workstations – This type of workstation is covered by two sides with an opening at one end that suits more for individual operations, and it allows more privacy.

Ergonomic Workstations – These workstations allow workers to stand up and stretch their bodies during the workday. Ergonomic workstations are designed to reduce physical strain and improve comfort and productivity.

Corner Workstations – These desks are a good option for workers who need multiple workstations for different tasks. They can also be a good choice for space-saving design. However, some workers may find it awkward to sit in the corner of two surfaces, and others may feel like the space behind the monitor is wasted. Corner workstations may not be the best choice for collaborative settings.

Cubicle Workstations – These workstations are a good option for workers who prefer to work alone in a quiet, focused environment. Cubicles can also be a good choice for large, open spaces because they can be arranged in many different configurations. However, cubicles may not offer the same level of privacy as individual offices, and the low walls can make it hard to concentrate due to environmental noise.

Benching – The benching plan is a wide-open office with little to no defined personal space. Workers are encouraged to use their laptops, find a spot to sit wherever they like, and get to work. It promotes communication, collaboration, and teamwork. There are no partitions or privacy. This arrangement can affect the worker's focus and productivity.

Privacy Cubicles – These are cubicles with a desk, chair, and four walls that give some privacy to the worker. They are designed for workers who value having personal space and require a dedicated workspace.

Linear Workstations – This type of workstation can be of different sizes and normally will have a partition in front of the desk with a designated height. In a linear workstation the workstations are generally arranged in a single row.

Modular Workstations – Modular workstations provide flexibility and customization. Modular systems often include interchangeable components, such as desks, partitions, and storage units, allowing the worker to create a workspace that suits their specific requirements. This style of workstation is particularly beneficial in dynamic office environments where layouts and team sizes may often change.

Team Stations – Team Stations are designed to have multiple workers within the same space. This type of station is ideal for those working on the same team enhancing communication and deliberation on business strategies without wasting time scheduling a conference room.

12.2.3 Factors to Consider for an Office Workstation

Ergonomics – Ergonomics ensures a comfortable workspace for employees. An ergonomic setup includes adjustable chairs, desks that can be raised or lowered, and monitor stands that keep employee screens at eye level. These adjustments keep the employee's posture aligned and reduce the risk of aches and pains.

Workstation Space and Layout – The size of the office should be considered when choosing a workstation. Measure the space to ensure the desk fits comfortably without crowding the room. Assess how the desk will fit with other furniture and whether it provides enough room for movement. A well-designed layout creates an organized and functional work environment.

Technology Requirements – Employee technology needs should be considered in designing a workstation such as the use of multiple devices or any specific tech requirements.

Budget – The budget is an important factor in choosing a workstation. The budget needs to balance the cost with the quality and features that are needed in a workstation. A good-quality workstation guarantees long-term comfort and productivity benefits.

Long-Term – Consider a workstation that will last and can adapt to changes.

Personalized Workstation – Employees should include things that make them happy and help them increase productivity, such as decorations, photos, or plants.

12.2.4 Workstation Safety

Workstation safety is important for preventing physical injuries and musculoskeletal disorders. Recommendations for creating a safe workstation:

- **Adjust the Workstation** – Ensure the desk and chair are adjustable, and that the worker's feet can rest flat on the floor. The monitor should be at eye level, and the worker should be able to look straight ahead without tilting their head.
- **Arrange Items within Reach** – Keep frequently used items like keyboard, mouse, and phone close by so the worker does not have to stretch or reach too far.
- **Take Breaks** – Workers should take regular breaks to stretch and vary their tasks.
- **Reduce Glare** – Use a combination of low-powered lamps to reduce glare from the worker keyboard and other surfaces. The worker can also tilt their monitor slightly and set it to dark text on a light background.
- **Reduce Noise** – Use earplugs, soft music, or a fan to reduce distracting noise.
- **Maintain a Comfortable Temperature** – Use layers of clothing or a fan to keep the temperature comfortable.
- **Training** – Make sure workers are trained in how to set up their workstation and recognize hazards.
- **Ergonomic Assessment** – Consider getting an ergonomic assessment if the worker is experiencing issues, has made adjustments to the workstation, or is new to the job

12.3 THE USE OF DRONES

Drones, often referred to as unmanned aerial or terrestrial vehicles, are aircraft or land vehicles that are operated by a ground-based operator. Unmanned Aircraft Systems (UAS) is a more general term that includes drones. UAS are often used for professional applications that require more complex functionalities and longer flight times than recreational drones. A drone's primary function is flight, but it may also have technology that allows it to record both audio and video.

The use of drones has increased considerably in recent years. These remote-controlled aircraft which vary in size are employed in a variety of applications by individuals, industries, and government. Drones can assess hazardous environments and collect data without placing workers at risk of injury. Drones may be used, for instance, to evaluate the air quality in confined spaces, and inspect mobile phone towers and rooftops.

Drone data collection technology has advanced significantly in recent years. It is common to find drones with 4K cameras with sophisticated sensors, artificial intelligence, thermal imaging, 3D, and light detection and ranging (LiDAR) cameras. Also, drones can be equipped with a range of sensors and transducers that sense various parameters including gas leaks and air pressures using electrical gravitational, thermal, mechanical, electromagnetic, and chemical capabilities. Drones can perform several surveying activities, many of which are currently being carried out commercially and professionally.[5]

Although OSHA does not currently have any regulations on the use of drones, the agency has sent regional administrators a memorandum outlining the protocol for utilizing drones when conducting enforcement operations.

People, businesses, and governments have used drones for a range of purposes, some of which are listed in Table 12.4 below.[6]

12.3.1 Drones Use in Improving Worker Safety and Health

Drones can be used to improve worker safety and health by reducing risk and worker exposure. Some of the ways that drones can be used are:

TABLE 12.4
Uses of Drones

Uses of Drones
Aerial photography
Express shipping and delivery
Disaster management
Search-and-rescue operations
Geographic mapping of inaccessible terrain
Building safety inspections
Precision crop monitoring
Unmanned cargo transport
Law enforcement and border control surveillance
Storm tracking and forecasting
Delivering medical equipment
Hazmat situations
Structural firefighting
Shore patrols
Tactical operations
Pursuit of criminals
Water rescues and dropping floatation devices

12.3.1.1 Reducing the Need for Working at Heights

Drones can perform inspections and surveys without the need for workers to, for example, climb ladders, scaffolding, or elevated platforms. This can help reduce the risk of serious injuries and fatalities.

12.3.1.2 Spotting Hazards

Drones can provide a panoramic view of a project, helping to identify hazards and monitor progress from a safe distance.

12.3.1.4 Inspecting Heavy Machinery

Drones can inspect heavy machinery and other equipment without putting workers in dangerous situations.

12.3.1.5 Training Employees

Drones can be used to demonstrate safe procedures and best practices without putting trainees at risk.

12.3.1.5 Streamlining Inspections and Reporting

Drones can streamline inspections and reporting, reducing the risk of human error.

12.3.1.6 Providing a Broader View for Safety Inspectors

Drones can provide safety inspectors with a broader view of the workplace.

12.3.1.7 Allowing Managers to View the Workplace Remotely

Managers can use drones to view the workplace from their office or remote area, allowing them to address issues that might not otherwise be identified.

12.3.2 Federal Aviation Administration

The current federal legislation does not directly address the use of drones in terms of privacy, and the Federal Aviation Authority's (FAA) drone regulations do not control or address privacy concerns. Many producers and users of drones are concerned about the formation of a regulatory framework of state and local laws, which will make it impossible for the sector to grow. States will most certainly have their drone regulations, and while federal regulations may be less stringent, public safety officials must understand and adhere to their state's laws. About 41 states have currently passed drone-related legislation, with an additional three states adopting resolutions.

The legislation addresses common issues such as defining what an unmanned aircraft system (UAS), unmanned aerial vehicle (UAV), or drone is, how law enforcement or other state agencies can utilize them, and how the general public can use them.

It is important to adhere to the FAA's rules that can significantly affect a public safety agency's drone program. The FAA now demands drone pilot certification for all flights in US airspace.

The FAA has tips and guidelines outlined on its webpage that include:

- **Line of Sight** – Keep your drone within the operator's visual line of sight.
- **Altitude Restriction** – Do not fly above 400 feet above ground.
- **Airport Restriction** – Never fly near airports.
- **Controlled Airspace** – Do not fly in controlled airspace without FAA authorization.
- **FAA Instructions** – Follow all FAA airspace restrictions, including special security instructions and temporary flight restrictions.
- **Aircraft Restriction** – Never fly near other aircraft.
- **Right of Way** – Always give way to all other aircraft. You are responsible for knowing your drone's altitude and location**.**
- **Flight Over People** – Never fly over groups of people, public events, or stadiums full of people.
- **Emergency Response** – Never fly near emergency response activities.
- **Conduct of Flight** – Never fly under the influence of drugs or alcohol.
- **Certification** – Take and pass an aeronautical knowledge and safety test and maintain proof of test passage to be made available to the FAA or law enforcement upon request.

12.4 VISUALIZATION TECHNOLOGY

Visualization is the process of developing a mental image of one's goals. By picturing how a specific event would look. Work may be done backward and created in smaller steps to make that goal a reality. Visualization technology uses graphics to help people understand data and has a wide range of applications, including:

- **Data Visualization** – Tools that use charts, graphs, maps, infographics, and even animations to assist users in seeing patterns and trends in data. These tools are useful for presenting data to nontechnical audiences and are essential for analyzing large amounts of data. The visual displays of information communicate complex data relationships and data-driven insights in a way that is easy to understand.
- **Process Visualization** – A way to present production processes and machine data using diagrams, charts, and curves. Process visualization is the process of transforming data into images to help visualize a process. It can provide a strategic overview of a process and improve workflow and the overall improvement of the process. Some methods that can be used for process visualization include:
- Business Process Modeling Notation (BPMN) – A standard language used across many organizations to visualize process workflows.
- Process Mapping – A workflow visualization technique that can help companies plan for process improvement. Process maps can be in the form of a process flowchart, functional flowchart, or process model.

- Flowcharts – Also known as process diagrams, flowcharts use a series of steps to represent a process. Each step in the flowchart represents a single action in the process.
- Data Visualization Tools – Some tools that can help with data visualization include IBM Cognos, Microsoft Power BI, Qlik, and Tableau.
- **Image Processing –** The use camera technology to automatically produce images that can help identify features of components. Image processing involves performing operations on an image to make it better or to retrieve important information.
- **Virtual Reality (VR) –** An immersive simulated environment that is created digitally and viewed inside a closed space. VR is fully immersive and replaces the real world with a simulated one. In VR, users are restricted to interacting with the virtual environment.

Augmented Reality (AR) – Can enhance objects and environments in the physical world with superimposed computer graphics. AR is partially immersive and adds digital elements to the real world. In AR, users can interact with both virtual and physical objects.

- **Mixed Reality (MR) –** Combines elements of VR and AR to create new visualizations.
- **Rea-Time Rendering (RTR) –** Can be used to quickly render games in real time as the player moves their character through an environment.

12.4.1 Visual Technology Used to Improve Workplace Safety

Visual technology can be effective in improving workplace safety through various means. Some of the means are discussed below.

Detecting Hazards – Computer vision can identify potential hazards, such as unattended equipment, machinery malfunctions, and obstructed pathways. It can also detect unsafe behaviors, like not wearing PPE or operating equipment unsafely.

Providing Real-Time Warnings – Computer vision can provide real-time alerts to employees and management to help prevent accidents.

Improving Compliance – Computer vision can help ensure that employees are consistently adhering to safety protocols.

Enhancing Productivity – Computer vision can streamline safety procedures and provide instant guidance to improve worker efficiency.

Facilitating Incident Investigations – Computer vision can help identify the root causes of incidents and accidents.

Enhancing Security – Computer vision can detect and alert authorities to potential security threats.

REFERENCES

1. Aksüt, G., Tamer, E. R. E. N., and Alakaş, H. M. Using wearable technological devices to improve workplace health and safety: An assessment on a sector base with multi-criteria decision-making methods. *Ain Shams Engineering Journal*, 15(2), 102423, 2024.
2. Jacobs, J. V., Hettinger, L. J., Huang, Y. H., Jeffries, S., Lesch, M. F., Simmons, L. A., and Willetts, J. L. Employee acceptance of wearable technology in the workplace. *Applied Ergonomics*, 78, 148–156, 2019.
3. Ma, J., Ma, D., Li, Z., and Kim, H. Effects of a workplace sit–stand desk intervention on health and productivity. *International Journal of Environmental Research and Public Health*, 18(21), 11604, 2021.
4. Garland, E., Watts, A., Doucette, J., Foley, M., Senerat, A., and Sanchez, S. Stand up to work: Assessing the health impact of adjustable workstations. *International Journal of Workplace Health Management*, 11(2), 85–95, 2018. https://doi.org/10.1108/IJWHM-10-2017-0078
5. Watkins, S., Burry, J., Mohamed, A., Marino, M., Prudden, S., Fisher, A., and Clothier, R. Ten questions concerning the use of drones in urban environments. *Building and Environment*, 167, 106458, 2020.
6. https://www.911.gov/assets/The-Uses-Continue-to-Emerge--Public-Safety-Drones-and-Considerations.pdf

13 Current Challenges for Safety and Health Workers

13.1 PROTECTING VULNERABLE WORKERS

Worker protection in the workplace is a primary responsibility placed on a company and its leadership team. Protection of workers can take on various forms depending on workplace activities, the hazards that may be present, and the ones created during the course of performing work. Various attributes can lead to a worker being placed in a vulnerable category and a worker that is considered vulnerable may not always be vulnerable. For example, a worker entering the workforce for the very first time may be vulnerable for a time period during which acclimation to the workplace is occurring. Once the worker has all of the necessary training needed to understand their job tasks and the knowledge to perform them safely, they may be considered less vulnerable. Another example for consideration involves females of childbearing age. When a woman has declared pregnancy, they may enter the vulnerability category especially if the worker is working in a high-hazard environment where radiation and chemicals are present. Once the pregnancy declared is over, the worker may no longer be considered vulnerable. Therefore, vulnerability among the workforce can change from time to time. Workplace laws and regulations is one avenue that can be used to facilitate protection of vulnerable employees while at work.

Vulnerable workers are those workers that are considered to have greater exposure than average to potential injuries and illness in the workplace. The workplace of today is supported by a segment of workers that may be classed as vulnerable. This category of workers may include:

- Workers that have physical and mental disabilities. These mental issues may or may not have resulted from the workplace. May require adjustments to work tasks or equipment
- Pregnant Women: Need accommodations to protect their health
- Workers that have limited knowledge and experience of the tasks that they have been hired to perform
- Workers that have different cultural experiences
- Workers that have a language barrier: Difficulty understanding safety instructions
- Workers new to the workforce: Lacking experience and familiarization with the job

DOI: 10.1201/9781032624389-17

- Workers that are overworked
- Migrant Workers: May face language barriers and lack understanding of local workplace rights
- Shift Workers (night shifts): Disrupted sleep patterns can affect alertness and safety
- Older Workers: Potential physical limitations and may be less adaptable to changing work demands
- Young Workers: New to the workforce and may lack awareness of safety protocols

As a note it is important to recognize the difference between vulnerable workers that can impact worker safety and health, productivity, and the vulnerable acts of workers that can be helpful in forming a workplace culture that will enhance safety, communication, and comradery needed to formulate teams that are productive. Some vulnerable acts that are helpful in forming and maintaining a safe and productive environment are listed below in Table 13.1 along with some potential benefits.

These acts of vulnerability are helpful in improving and facilitating a balanced workplace where improvement can be continuous as well as provide improvement in the culture of an organization. The point is that vulnerability in the workplace depending on the action or situation can also be beneficial. However, there are times

TABLE 13.1
Potential Acts of Vulnerability

Vulnerable Actions	Workplace Benefits
Authenticity	Expressing freely and openly one's feelings, personality, and opinions. Self-expression ensures you make and voice decisions you can support that align with your belief
Requesting assistance	Admitting when assistance is needed thereby admitting a lack of knowledge. Workers must feel comfortable in requesting assistance when needed
Admitting mistakes	Mistakes should be viewed as teachable moments and not an opportunity to assign blame
Admitting a lack of knowledge or understanding	Workers must feel comfortable in admitting when there is a lack of knowledge or clarification is needed
Interacting with coworkers	Demonstrating trust while interacting with coworkers and management
Receiving constructive feedback	Feedback can be positively or negatively viewed. When feedback is negatively viewed it may create a culture where workers are not open to asking questions or using knowledge effectively
Offering a different point of view	Differing points of view that are dismissed can lead to information being withheld and therefore loss of opportunity to correct mistakes before they become issues
Challenging the way work is performed	Challenging current practices can lead to improved practices and improved productivity

and groups of individuals that are vulnerable and as a result of their vulnerabilities, they can be taken advantage of, cast aside, or become injured if not cared for during the course of performing work. Some of these groups are discussed in Sections 13.1.1–13.1.8 to provide more context on their vulnerabilities, the laws that have been enacted to protect them, and the employer's responsibilities to protect them from workplace injuries.

13.1.1 Workers with Disabilities

Many workers who enter the work environment daily are classified as being disabled according to the definition of having a disability by the Americans with Disability Act (ADA). The ADA defines a person having a disability as one having a physical or mental impairment that limits one or more life activities substantially. This federal civil rights law has and continues to protect those workers that are disabled from entering and functioning in the workplace. This law is broad and far reaching and covers many aspects of accessibility for people that have a disability.[1] Employers are obligated to provide people with disabilities with equal opportunities to benefit from employment opportunities that are available to others.

Workers can become disabled after entering the workforce. In fact, a worker can become disabled as a result of their work experiences or accidents that occurred as a result of performing their job responsibilities. These workers can be vulnerable depending on the type of work being performed and the work environment. Having a disability does not mean that a worker is vulnerable; however, they can become vulnerable or more vulnerable based on the work environment.

13.1.2 Pregnant Women

The workforce of today consists of a large percentage of women. In fact, the Bureau of Labor Statistics (BLS) reported that in 2020 women made up 56.1% and in 2021 56.8% of the workforce.[2] This trend has been growing over the years and decades and is not expected to slow. Single-parent households with mothers as the head of the household continue to increase. Women are working in various roles in the work environment from office work, medical, service, hospitality, to construction.

By law, employers are required to provide reasonable accommodation to women while pregnant that will afford them the opportunity to continue working during their pregnancy term. Providing reasonable accommodation is easier in some workplaces and harder in others. Regardless of any complexity, these accommodations must be provided according to the Pregnant Worker Fairness Act (PWFA) of 2023. Although the PWFA only covers accommodations other laws that are enforceable make it illegal to discriminate against women during pregnancy or fire a woman on the basis of being pregnant and any related medical conditions.[3]

During pregnancy, women can be vulnerable to many workplace conditions and must be protected against being placed in environments that can harm them or their unborn child. For example, females that work in the chemical or nuclear industry have a greater risk of exposure during pregnancy to hazards from various chemicals and radiation emitting devices, therefore may need to be completely removed from those environments and placed in other settings such as in an office setting during pregnancy.

13.1.3 Workers Who Have Limited Knowledge and Experience

Knowledge in the workplace is inclusive of the awareness and the know-how that is necessary to solve problems in organizations. Some workers enter their work environment with the knowledge to immediately perform many of the tasks that may be required of them. The knowledge they bring to the workplace results from prior formal training and experience. Knowledge is not automatic, and workers can be trained to gain what's needed to maneuver the workplace safely; contribute to the success and sustainability of an organization; and become a viable part of their environment.

The workplace contains workers of varying knowledge levels and experience. The less experienced workers are typically vulnerable to accidents and making mistakes that can impact their health and safety and the health and safety of their coworkers. Productivity and product quality often suffer as a result of employees' lack in knowledge. It is incumbent upon management to acclimate workers to the workplace through avenues such as training (classroom and on the job) and providing a buddy or mentor to assist with guiding them as they perform their work. In addition, management must make themselves available, observe work performance, and make corrections as necessary.

13.1.4 Workers That Have Different Cultural Experiences

The United States is recognized as a nation of immigrants, and as a nation of immigrants, the culture of an organization can become more complex and difficult to facilitate into an integrated system to support organization success. Therefore, management is placing more focus on the culture of their organizations as they continue to realize the importance of culture. The culture of an organization can serve as the foundation of the organization that facilitates how work is performed, the interaction between members and customers, and enhances the ability for members to build trusting relationships. Human performance is significantly dependent on culture as culture has an influence on members in terms of their willingness to release creativity and support the vision of the organization. Important components of a viable culture include:

- Employees that are engaged
- A vision and a mission that is supported by members of the organization
- Values that are clear, sound, and communicated that document the guidelines for expected behaviors
- Sound practices that demonstrate value for their employees and customers and a leadership team that "walk the talk"

Employees coming together from various cultures not only can be vulnerable, but they can also increase vulnerability and risk for other workers working alongside them. For example, a new employee coming from an environment that did not place value on completing work safely may not utilize the safety training received during onboarding. During the performance of their work tasks, they may take shortcuts that create an unsafe work environment for themselves and coworkers. It is not easy to activate the components of a sound culture when the workforce is diverse having varied cultural experiences. Although complex, management must work toward an inclusive culture.

13.1.5 Workers Having a Language Barrier

Communication in the workplace is crucial in getting work accomplished and developing a safety culture that is protective of employees as they interact. Language is the fundamental way people communicate, and language barriers can pose significant challenges affecting the way business is transpired to include safety of the workplace. Barriers to communication can create issues such as:

- Confusing instructions provided to workers
- Workers having difficulty in understanding instructions
- Increase safety risk
- Reduction in productivity
- Reduction in product or service quality
- Frustrations among employees

Language barriers introduce a level of vulnerability and restrictions that can cause misunderstanding and restrict the use of knowledge. The resultant vulnerability impacts not only the worker experiencing difficulty in communication but also coworkers and the entire organization. To overcome and reduce this barrier management can provide tools such as language classes, hire a translator, use visual aids, and publish information in more than one language.

13.1.6 Workers New to the Workforce

New workers entering the workforce is a constant activity necessary to keep businesses afloat. On the other hand entering the workforce is an expected event for people to earn funds to financially secure their present and future. New workers do not necessarily have the skills needed to perform the tasks they have been hired to perform or have the skills and knowledge to contribute to productivity and work safely. Statistics have shown that employees that are new to the workforce are more likely to become injured. Perhaps it's partially because these workers are eager to show their manager and coworkers what they know and that they can quickly become an integral part of the team.

To reduce worker vulnerability and the risk associated with workers new to the workforce, many companies have robust onboarding processes that include training from human resources to safety. In high-hazard environments, hazard analysis training can serve as an avenue to acclimate new workers and assist in helping them gain confidence in their ability to perform their work tasks safely. Other avenues that should be considered to reduce new worker vulnerability include a competent leadership team, management having an open-door policy, and encouraging open and honest communication in all interactions.

13.1.7 Overworked Workers

In the current environment, it is hard not to recognize that some companies or merchants are operating with less staffing that is required to serve the needs of their customers and to improve efficiency and productivity for the business. These businesses are operating in the customer service industry to the manufacturing industry. The lack or shortage of workers has created wait-time for customers, inability to get products needed, or the ability to receive products and services in a timely manner. In addition to creating complexity for customers, the stress placed on workers is great to the point that some workers quit and seek employment with other companies or leave the workforce completely. The stress created for these employees can result from the expectation for workers to work additional hours or work long hours impacting their work-life balance. A worker who has reached their limits in terms of productivity and does not have the necessary time to regenerate/recharge before returning to work in most cases is overworked.

Workers that are stressed and are not rested render themselves vulnerable to providing poor customer service, producing poor quality products, and prone to workplace injuries to themselves and coworkers. Workers that are overworked typically may not communicate to management that they are feeling stressed concerning the amount of work expectations for fear of being labeled as being a complainer or even lazy. Therefore, these workers will come to work daily with apprehension and perform their work tasks and may not have their mind in the game. They are watching the clock distracted looking forward to the end of the workday. Some signs that an employee is overworked and unable to recharge prior to the next workday include the signs that are listed in Figure 13.1.

Management must be in touch with the demands of their business and the stress that is placed on workers to meet goals and produce services. Engagement in work activities and monitoring work is a key element in recognizing when a worker is being asked to produce above their capacity. Employers must continually evaluate their work strategies, balance them against the workforce, and determine what is feasible to accomplish while not overextending the workers. Some tips to prevent worker burnout are listed in Figure 13.2.

13.1.8 Migrant Workers

Migrant workers sometimes are referred to as seasonal workers and are workers who travel from one country to another to work. Migrant workers generally do not have the intention of remaining in the country in which they travel with the express

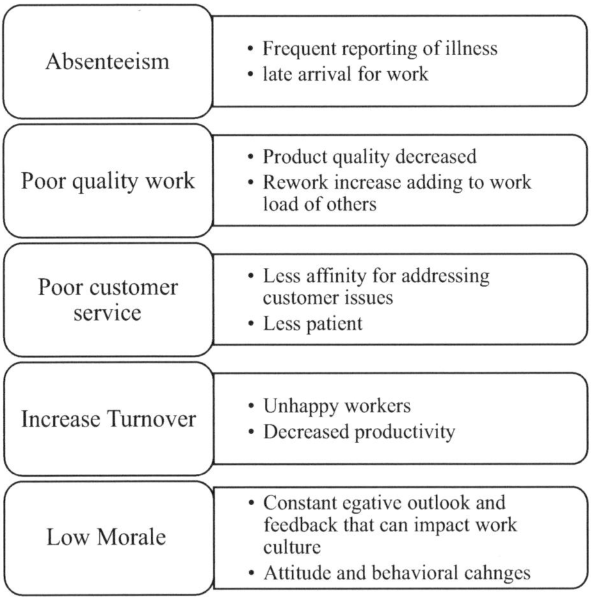

FIGURE 13.1 Signs of overworked employees

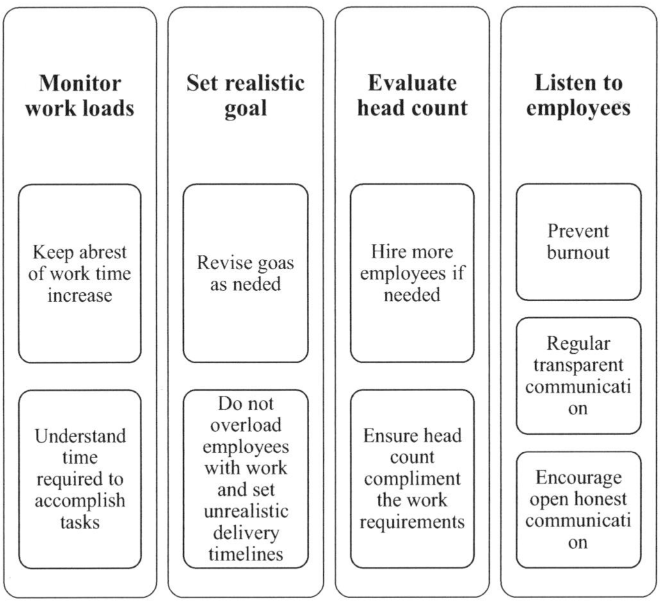

FIGURE 13.2 Tip to prevent worker burnout

purpose of gaining work opportunities. The reason for migrant workers seeking employment in another country is generally due to the lack of economic opportunities for these workers in their country of origin.

To protect these workers, the Migrant and Seasonal Agricultural Worker Protection Act (MSPA) was enacted in the United States. The Act establishes employment standards related to wages, recordkeeping, housing, transportation, disclosures, and recordkeeping. Although MSPA is designed to protect the rights of an important vulnerable class of workers, reports of unfair treatment are still reported, but not as prevalent as it has been in prior years and decades. Other countries have similar laws to protect these workers as the work they perform is valuable to aid in the sustainment and growth of the economies in many countries.

13.2 RECOGNIZING AND PREVENTING WORKPLACE VIOLENCE

Workplace violence has been occurring more frequently within the past decade prompting companies to train workers in ways to recognize potential adverse behaviors, protect themselves, and report as soon as they become aware of violence or the potential for violence to occur. Many workplaces require annual training on workplace violence and have policies indicating their position with regard to violence and the measures taken when workers engage in such behavior. Workplace violence is considered acts or threats of physical violence, harassment, intimidation, or other threatening behavior that occurs in the work environment. This type of violence can range from threats, verbal abuse, or physical assaults and even homicide. Each year many workers report having been victims of workplace violence or knowing someone who has been a victim. Although many cases are reported there are many cases that go unreported. Therefore, it is not feasible to get an accurate account of how serious the issue of violence is in the workplace. Some factors that can increase the risk of a person being subjected to workplace violence include:

- Exchanging money with the public
- Working with volatile people
- Working with unstable people
- Working alone or in isolated areas
- Working where alcohol is served
- Working late at night

One of the best protections employers can offer their employees is to establish and enforce a zero-tolerance policy with regard to workplace violence. In addition, assessing the workplace at some frequency may be helpful in identifying workplace violence risk factors. Workplace violence is a growing concern for employers and employees across the globe. More than two million workers are vulnerable to becoming victims of workplace violence yearly. Although workplace violence can occur anywhere at any time, some workers are at increased risk. According to OSHA, nearly 2 million workers in the United States are victims of workplace violence each year. Some industries with a higher risk of fatal violence include sales, protective services, and transportation. Healthcare and social assistance workers have a higher risk of nonfatal violence that results in days away from work.[4]

13.3 MAINTAINING WORKPLACE EQUITY

Workplace equity is at the forefront of creating and maintaining an inclusive work environment. Equity is achieved by closing the gender and racial gaps between pay and advancement opportunities for workers. Equity in the workplace has been a long-standing challenge that many companies have attempted to provide for their staff. Although work is being done to ensure equity, the issue still exists in many workplaces. The lack of equity in the workplace is a reason why some workplaces experience frequent turnover and in turn experience issues in filling open positions. Employees who exit a workplace because of perceived inequity often share their perceptions and experiences with others in their sphere of influence. This sphere is far reaching thanks to social media and other networking avenues and platforms.

Organizations that believe in and are committed to an equitable workplace for all workers take a proactive approach to prevent and address unfairness in the workplace. These organizations train their managers on how to recognize inequities and mitigate them through policies, procedures, and actions. Some actions that can be taken to facilitate equity within organizations include:

- Conducting an equity audit using a reputable third-party firm
- Monitoring organizational culture and ensuring equity expectations. Policies and procedures are embedded in the culture
- Providing training on diversity, inclusion, and unconscious bias
- Recognizing when policies and procedures are not being followed or not effective in facilitating an equitable culture
- Facilitating and engaging in open and honest communication about equity concerns
- Facilitating equitable recruitment, hiring, and promotion

13.4 WORKPLACE LAWS AND REGULATIONS

Many laws and regulations are written to protect the safety and rights of employees while in the workplace. These laws and regulations were made necessary throughout the decades resulting from issues that workers had endured in the workplace. These issues in many cases resulted in harsh treatment of employees and employees suffered with little or no recourse to solving these issues to their benefit. The harsh treatment of workers formed the basis for many of the laws written. For example, the Occupational Safety Administration was created in 1970 by Congress as a result of public pressure over the number of injuries, illnesses, and fatalities that were occurring in the workplace.

Although laws and regulations are developed and enforced, not all workplaces are in compliance with these laws and regulations. A lack of compliance adds to the number of employees injured in the workplace daily. Less serious injuries go unreported and therefore are not included in the injuries that are accounted for by each workplace annually.

Employers should strive to create a culture that supports a positive work attitude among workers, facilitates trust, and fosters productivity and loyalty. Included in a positive culture is ensuring that laws and regulations are followed and monitored to ensure compliance.

13.4.1 Occupational Safety and Health

The Occupational Safety and Health Administration (OSHA) is a regulatory agency within the Department of Labor (DOL) that focuses on assuring safe and healthful working conditions in the workplace. Under OSHA employers have the responsibility to provide a safe workplace for workers. Simply put, some of the requirements under OSHA state that:

- Employers must provide workers with a workplace that does not have serious hazards
- Employers must follow all OSH Act safety and health standards
- Employers have a responsibility to identify and resolve safety and health issues in the workplace
- Employers must attempt first to eliminate or reduce hazards by making changes in working conditions as opposed to relying on protective equipment
- Employers must inform workers about chemical hazards located in their work environment

The OSHA also provided rights for employees (full-time and temporary employees) with regard to having the ability to work in a workplace that is relatively hazard free and does not impose a risk of serious harm. Table 13.2 lists some of the standards that are promulgated under the OSHA to help maintain workplace safety.

The standards that make up the OSHA Act are a key component in reducing worker vulnerability in the workplace and ensuring compliance with the Act.

13.4.2 Department of Labor

The Department of Labor (DOL) has responsibility for administering federal labor laws to guarantee workers' rights to a safe, fair, and healthy work environment. This guarantee ranges from minimum hourly wage, overtime pay, protection against employment discrimination, and unemployment insurance. The various laws administered by the DOL to ensure employee vulnerability is eradicated are shown in Figure 13.3 and defined in Table 13.2.

Table 13.3 outlines in brief workplace laws and the type of protection it seek to provide to workers as they perform the job functions in the various workplaces.

TABLE 13.2
Vulnerability Limiting Regulatory Requirements

Standard 1910	Potential Worker Vulnerability	Purpose
Subpart D	Walking-Working Surfaces	Focused protection of employees from hazards that may result from walking-working surfaces in the workplace
Subpart E	Exit Routes and Emergency Planning	Establishment of the minimum requirements for providing a reasonable level of life safety and property protection from hazards such as fire or dangerous conditions in buildings
Subpart F	Powered Platforms, Manlifts, and Vehicle-Mounted Work Platforms	Provide protection for workers when working at heights and using special equipment to provide access to the work to be performed
Subpart H	Hazardous Materials	Provides requirements for hazardous materials such as compressed gases, flammable liquids, explosives, and blasting agents
Subpart I	Personal Protective Equipment	Set guidelines for the use of personal protective equipment in workplaces to include maintenance, hazard assessment, and training
Subpart L	Fire Protection	Establish requirements for fire brigades, fire suppression equipment, fire detection systems, and fire alarm systems to aid employers in providing a safe and healthy workplace for their employees
Subpart Z	Toxic and Hazardous Substances	Establishes requirements to prevent employee exposure to toxic and hazardous substances

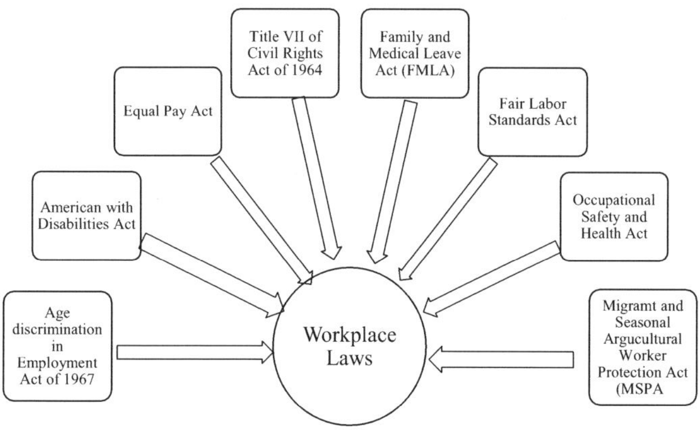

FIGURE 13.3 Workplace laws[5]

TABLE 13.3
DOL Workplace Protection[6]

Labor Law	Protection Offered
Age Discrimination in Employment Act of 1967	Prohibits employment age discrimination and prevents older workers from being disadvantaged in their efforts in finding and retaining employment
American with Disabilities Act	Provide protection for people with disabilities as it pertains to their public life. The Act prohibits discrimination based on disability in employment, public and government accommodations and facilities, telecommunications, and transportation
Equal Pay Act	Sets the requirement that men and women in the workplace receive equal pay for equal work
Title VII of Civil Rights Act of 1964	Designed to protect employees and job applicants from employment discrimination based their sex, race, color, religion, and national origin
Family and Medical Leave Act (FMLA)	Provides eligible employees with unpaid job-protected time off. During this leave, health insurance benefit is maintained, and employees are entitled to return to their same or equivalent position at the end of the leave
Fair Labor Standards Act	Establishes minimum wage, overtime pay, recordkeeping, and standards for employing youth. These requirements impact employees in Federal, State, and local governments as well as private sectors
Occupational Safety and Health Act	Support employers providing safe and healthy workplace for workers
Migrant and Seasonal Agricultural Worker Protection Act (MSPA)	Provide protection for migrant and seasonal workers through establishing employment standards related to wages, housing, transportation, and recordkeeping

REFERENCES

1. https://www.ada.gov/topics/intro-to-ada/
2. https://www.bls.gov/opub/reports/womens-databook/2022/home.htm#:~:text=Selected%20demographic%20characteristics,in%202020%20(56.2%20percent
3. https://www.eeoc.gov/wysk/what-you-should-know-about-pregnant-workers-fairness-act
4. https://www.osha.gov/workplace-violence
5. https://www.usa.gov/workplace-laws, Workplace Laws
6. Workplace laws | USAGov

14 Barriers to Safety and Health

14.1 SAFETY AND HEALTH PROGRAM

In an organization, it is critical to maintain effective health and safety standards. It is necessary to safeguard the well-being and safety of employees, customers, and the public. However, there are often barriers that can hinder the achievement of these standards. The objective of the safety and health practitioner along with management is to overcome these barriers to attain high health and safety levels. To overcome occupational health and safety barriers, a complete strategy is required, including risk identification, recognition, control, and management and employee engagement.

Certain jobs or processes can be inherently complex, involving multiple steps, hazardous materials, equipment, or hazards. This complexity can make identifying hazards challenging. The complexity of some exposures such as chemical exposure lies in managing and minimizing exposure through effective ventilation systems, personal protective equipment (PPE), and frequent monitoring. The complex nature of some work processes can make it challenging to implement and maintain adequate safety measures, including proper ventilation, enclosure, and emergency evacuation plans.

Operations related to safety and health are sometimes hampered by conflicting goals within an organization. These conflicts may develop between the need for safety requirements, budgetary restrictions, and production targets. The significance of health and safety can be overshadowed by the pressure to fulfill production targets or deadlines, which can result in standards being compromised. Overwork, pressure to meet deadlines, and time constraints might result in taking shortcuts or disregarding safety precautions. When workers are overworked, they may put production ahead of safety, which would lower safety standards.

Ten factors that can create barriers to achieving effective health and safety standards are shown in Figure 14.1 and discussed in further detail in Sections 14.11–14.10.

14.1.1 Lack of Employee Engagement

A major barrier to the success of health and safety programs is low employee involvement. The development of a safety-conscious work culture is hampered when workers do not believe that they are actively participating in safety efforts or programs. This barrier may be addressed by promoting open communication, soliciting employee feedback, and including staff members in the decision-making process. A work environment is created where everyone feels accountable for their well-being

DOI: 10.1201/9781032624389-18

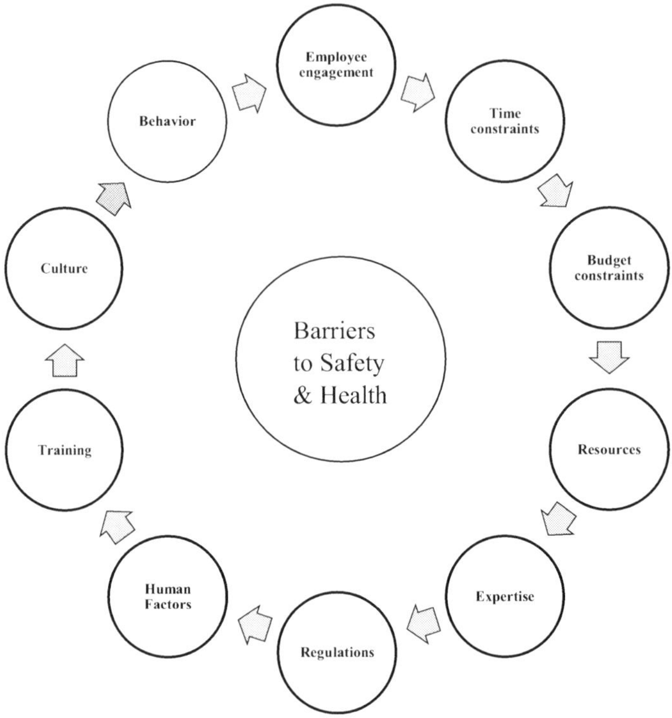

FIGURE 14.1 Barriers to safety and health

by encouraging a collaborative mindset and giving employees the freedom to take charge of their safety is necessary in the creation of a healthy safety culture.

14.1.2 Time Constraints

When there is limited time available for an organization to prioritize and implement health and safety measures due to competing demands for productivity or meeting deadlines, focus on health and safety often weakens and employees are potentially at risk of occupational exposure.

14.1.3 Budget Constraints

Budgetary conflicts can occur when there is insufficient financial allocation for organizational health and safety programs leading to disparities between cost considerations and the investment required for adequate safety measures. Organizations during economic downturns tend to prioritize cost-cutting measures leading to the reduction of budget allocation to health and safety programs. Underfunding of health and safety programs can have serious consequences, resulting in a rise in accidents and injuries to long-term health problems for employees.

14.1.4 Inadequate Resources

It is often difficult to maintain a safe working environment when the required resources are not available. Inadequate infrastructure and resources can be a major barrier to the implementation of safe and healthy practices. This includes using outdated equipment, inadequate safety gear, poor maintenance, and a lack of safety signs. Implementing appropriate safety measures may be hampered by a lack of equipment, including personal protection equipment (PPE), safety tools, and maintenance equipment.

14.1.5 Expertise

The absence of specialized knowledge or abilities in health and safety management might create an expertise barrier that impedes the efficient execution and maintenance of safety regulations and exposure management and surveillance.

14.1.6 Different Regulations

Legislation or standards within an industry or across various jurisdictions can vary. International businesses that conduct business across borders may find it difficult to understand and abide by the various health and safety laws. The implementation of consistent safety measures might be complicated by variations in safety standards, regulations, or cultural norms in different locations, which may result in discrepancies and possible gaps in compliance.

14.1.7 Human Factors

The interaction that exists between workers, tasks, and their surroundings that have the potential to affect their health or safety are referred to as human factors. These elements can lead to human error and accidents and include distraction, fatigue, stress, and insufficient training. Preventing mishaps in the workplace requires an understanding of and attention to the human aspects. By implementing fatigue management techniques into practice, offering sufficient rest periods, and cultivating a positive work atmosphere that promotes candid conversations about mental health, organizations can work toward overcoming this obstacle. In addition to improving skills and abilities, frequent training and continuous professional development can lower the risk of errors brought on by inadequate training and lack of focus

14.1.8 Lack of Training

Inadequate training and lack of awareness are two major obstacles to implementing successful health and safety programs. Many workers may be unaware of all the possible risks in their workplace or the appropriate safety protocols to adhere to. When workers do not know about the possible risks and hazards, they may not adequately prioritize safety precautions. Accidents and risks might increase because of inadequate training.

14.1.9 Poor Safety Culture and Practices

When organizations fail to provide adequate resources for safety programs, they may unintentionally discourage safe behavior. Employees' motivation to put safety first may be weakened if they believe that others are being rewarded for risky behavior and that safety is not given enough attention. Recognizing and rewarding appropriate behavior is a great contributor to enforcing the desired work practices and behaviors that will enhance the culture of the organization.

14.1.10 Worker Behavior

A positive health and safety practice often relies on the positive behavior of the worker, and sometimes workers do not behave in the work environment the way they were trained. Workers' behavior can be a significant barrier to maintaining good standards in the workplace.

14.2 EXPOSURE SURVEILLANCE

Surveillance is a core activity in developing and ensuring an effective safety and health program that is able to respond to the changes in regulatory requirements and protect workers from hazards. The two primary types of surveillance that comprise an effective exposure control program include hazard and health surveillance. Both types of surveillance (Figure 14.1) are most effective when implemented in concert and have significant roles in the overall exposure surveillance program for an organization.

An effective hazard surveillance program should identify risks and result in the reduction of workplace exposures, which will eventually reduce the occurrence of injuries, illness, and work-related diseases. In addition, exposure surveillance provides an opportunity to monitor trends and observe emerging patterns of workplace hazards which can be used to predict future illness and injuries. Health surveillance is another key element in the exposure assessment protocol. It can take the form of clinical or physiological assessment of a worker or workgroup. When observing individual exposures, the goal is to detect adverse health effects resulting from occupational exposures early on, so that appropriate preventive measures can be instituted promptly removing the worker from the hazardous environment.[1]

There are barriers to an effective health and safety program that can have a significant impact on exposure surveillance. For example, an exposure surveillance program will not be successful if it is not implemented by professionals who have the expertise and knowledge in monitoring and data analysis. Typically, exposure surveillance is performed by individuals trained in safety and health specifically in the area of industrial hygiene. The industrial hygiene professional's primary focus is to protect workers from routine and unique hazards in the workplace. These hazards

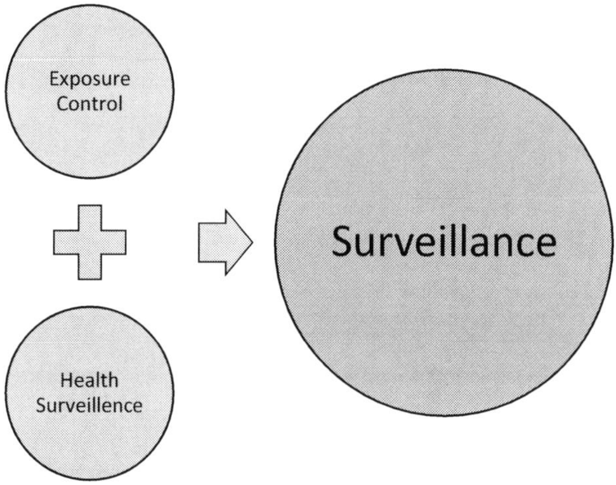

FIGURE 14.2 Comprehensive surveillance components

of concern generally fall into the category of physical, chemical, biological, or radiological hazards. These professionals are trained in the art and science of industrial hygiene, which makes them uniquely qualified to manage programs that are designed to protect workers while in the workplace. Industrial hygienists are taught various methods and techniques on how to recognize and manage occupational hazards.

It is not always feasible to attract, hire, and retain the services of these Subject Matter Experts (SME) because of the market need for their expertise and the shortage of professionals in the field. In addition, these individuals, because of their expertise, often demand pay that may not be considered comparable to the pay structure of an organization. Also, key in the exposure surveillance process is the expertise of the occupational medicine staff. Physicians trained in occupational medicine are not always available to work with organizations to evaluate the link between exposure and health conditions. The exposure surveillance process will not be effective if not managed by professionals who have the expertise to conduct hazard analysis, make decisions that ensure regulatory compliance, conduct monitoring, evaluate data, and work effectively with occupational medical providers. Table 14.1 provides additional impact of the barriers to implementing an effective exposure surveillance program. The purpose of implementing an effective exposure surveillance program is to protect workers from workplace hazards that can cause injuries or illness and reduce the associated risks.

14.2.1 Exposure Hazard Monitoring

At the core of a good surveillance program is exposure monitoring of the work environment and activities that take place within an organization. A primary characteristic of an effective exposure monitoring program is having the ability to identify, evaluate, and determine overall health risks to the worker from exposures to

TABLE 14.1
Barrier Impact on Exposure Surveillance

Barrier	Barrier Impact Examples
Lack of Employee Engagement	An important aspect of exposure surveillance is performing personal air monitoring of workers while performing work
Time Constraint	Difficulty in monitoring work tasks that are short in duration
Budget Constraint	Unable to hire skilled professionals to conduct surveillance, equipment needed, and affordability of the use of an accredited laboratory
Inadequate Resources	Resources including people and equipment needed to conduct monitoring unavailable
Expertise	The staff with expertise and knowledge in proper monitoring is not available
Different Regulations	Regulatory changes that facilitate change and update in surveillance. For example, lowering of an Occupational Exposure Limit (OEL)
Human Factors	Worker unwilling to wear monitoring equipment
Lack of training	Lack of knowledge or training in the proper surveillance method to use by the person conducting the surveillance and lack of training for the staff wearing monitoring equipment on proper handling of equipment
Poor Safety Culture and Practices	Safety culture does not support the importance of surveillance and protecting the worker and heavy reliance on Personal Protective Equipment (PPE)
Work Behavior	Worker's reluctance in supporting surveillance for example refusing to maintain

physical, chemical, biological, and radiological contaminants. The applicability and importance in determining worker exposure applies across all work activities and environments in the workplace, whether an office worker is working on the computer in the office, a chemical operator adding chemicals to the process, or a mechanic is performing preventive maintenance on equipment.

The potential exposure to hazardous contaminants and work environments, and subsequent health risk, is required to be evaluated and documented. Below are some reasons why exposure monitoring should be performed:

- Occupational safety and health regulations require exposure monitoring to be performed, and compliance with applicable standards must be demonstrated
- Engineering controls should be verified
- Workers have a right to understand what health risks they may be exposed to when performing their job functions
- Management has a responsibility to understand health hazards and risks created in the manufacturing of their product
- The selection of personal protective equipment must be evaluated, determined, and technically justified
- Provide assurance to workers that the workplace is safe
- Use exposure monitoring data to investigate potential exposure-related incidents or complaints.

Evaluating data from monitoring, understanding, and characterizing exposures resulting from conducting monitoring is key in assessing and reducing the risk of exposure to various chemical and biological constituents. When evaluating data, considerations should be given to the exposed population, the task being performed, toxicity and health risks, occupational exposure limit, and exposure time.

14.2.2 Health Monitoring

Health monitoring and surveillance of the worker encompasses the medical techniques, and the methods used to evaluate workers exposed to physical, biological, chemical, and radiological hazards. A comprehensive health monitoring program includes the specified frequency and program that validates that the workplace environment does not cause harm to workers.

A comprehensive health monitoring and surveillance program is an essential element of an overall worker injury and illness prevention program. The benefits of a health monitoring and surveillance program include the following:

- Ensuring that personnel are capable and able to perform a job prior to being assigned
- Reduce the cost of a company's health insurance
- Confirm to employees that the work they are performing does not impact their personal health
- Provide employees with access to wellness and injury illness prevention information
- Confirm that the hazard controls that are in place are successful and are not causing additional harm to the employees
- Comply with regulations that require the workplace to have a medical monitoring and surveillance program

A health monitoring and surveillance program must be tailored to the hazards and contaminants to which there is an exposure potential. There are common regulatory requirements associated with health monitoring such as requirements associated with asbestos, lead, chromium, noise, and heat stress. There are also regulatory requirements associated with health monitoring that are specific to various types of job functions, such as nuclear operations and remediation and hazardous waste site cleanup.[2]

14.3 IMPROVING EXPOSURE AND RISK ASSESSMENT

Exposure and risk assessment go hand in hand. Improving the risk assessment process used will invariability improve the exposure assessment process. Exposure risks may arise from a single contaminant or can be from several contaminants, which together represent an even greater health risk (synergistic effects). Improving worker risk and exposure to potential hazards begins with having a comprehensive company policy supporting and mandating a safe work environment, and developing and implementing a comprehensive risk assessment process that has the backing of management and funding to implement effectively. There are two basic methods used to conduct workplace risk assessments. The two methods are to either a quantitative or qualitative approach to collecting and evaluating data. Both methods are complimentary and needed to ensure accuracy of the data, proper quantification of risk, and development of risk mitigation measures. The preferred method of the two is to conduct exposure or risk assessments using a quantitative approach however, or a combination of the two provides additional information that can enhance information and knowledge gained.

The quantitative approach to determining health risk uses data collected through personal exposure monitoring, ventilation or air dispersion modeling, process monitoring, and area monitoring. Data that can be included in the quantitative approach can include chemical inventory, process data (run sheets, ventilation air exchanges), occupational exposure limits, personal monitoring data, and area monitoring data. The use of quantitative methods can be challenging since many times the necessary data is not available to complete the assessments.

There are times when data cannot be obtained to quantify concentrations of contaminants when performing exposure or risk assessments. Therefore, the need arises to conduct a risk assessment using only qualitative data. A qualitative risk assessment is based on the use of information, such as process-based, observed workplace conditions, and professional judgment. Components of a qualitative risk assessment can include chemical inventory, injury and illness incidents, regulatory requirements, cultural norms, assessment and inspection findings, and employee feedback. When the qualitative approach to characterizing and monitoring exposures is used, conservative decision-making must be applied (i.e., safety factors) to account for uncertainties in the risk identification and evaluation process.

Barriers that can have a significant impact on improving exposures and risks assessments are listed in Table 14.2. It is not necessary for all barriers to be evaluated at once that can stifle improvement. Everyone can contribute to moving a marginal program to a good program or a great program to a greater program as the plight for continuous improvement continues.

Some techniques to consider for Improving Exposure and Risk Assessment:

- Conduct a baseline risk assessment at the onset of a task or project
- Review risk assessment strategy at some periodicity (annually)
- Review exposure assessments at least every three years and update as soon as changes to the process or work tasks are made
- Ensure that exposure and risk assessments are performed by knowledgeable professionals with credibility in their ability to deliver a credible assessment

14.4 CERTIFYING EXPOSURE CONTROL

Certifying exposure control is generally accomplished using various technologies and equipment. The certification process is not an easy task, and it should begin with a plan detailing the technologies, equipment, and the process that will be used to

TABLE 14.2
Barrier Impact on Exposure and Risk Assessment Improvement

Barrier	Barrier Impact
Lack of Employee Engagement	Employees not engaged in hazard analysis
Time Constraint	Professionals having several deliverables and job responsibilities have limited time to devote to exposure and risk plan and improvement
Budget Constraint	In an environment where budgets are tight and spread across many programs and processes, the first tendency for some is to reduce support provided for some safety and health measures in favor of production
Inadequate Resources	Professionals are spread thin with limited ability to focus on exposure and risk improvement
Expertise	Limited or unavailability of knowledgeable person to develop, champion, and implement improvements
Different Regulations	Changes in regulatory requirements and implementation of newly promulgated ones
Human Factors	Lack of support for change
Lack of training	Individuals developing and implementing the exposure and risk assessment plan lack the knowledge to develop and implement the plan
Poor Safety Culture and Practices	Safety culture does not lend itself to open and honest communication that can lead to discovery of data to inform change and improvement
Work Behavior	Does not support changes that may be required to achieve improvement

complete the certification. In addition, certification should be completed by professionals having the training and knowledge to complete and document the certification process and results. Certification of exposure is a multifaceted activity that may involve the following at a minimum:

- Collecting extensive amounts of data
- Analyzing the data collected using statistical methods
- Comparing data results against the occupational exposure limits if one exists
- Carefully evaluate the tasks being performed and the environmental condition in which the task is being performed
- Ensuring that the certifier is skilled and has the experience and credibility and have the resources needed to certify the exposure and controls that will be used
- Review and understand the occupational standard that apply to the constituent being monitored

14.4.1 Exposure Control Plan

Ensuring exposures are controlled in the workplace starts with a plan that details the types of potential exposures that may be present and the methods that will be used to effectively control those exposures. Prior to embarking upon the effort to develop and implement an exposure control plan an effective exposure assessment process must be implemented. The exposure assessment process is the method used to evaluate and quantify contaminants and physical conditions that can cause health problems.

Key aspects of the exposure assessment process include:

- Defining workplace characterization and monitoring strategy
- Collecting data
- Data evaluation and determination of acceptability
- Documentation and communication of characterization and exposure data
- Reassessment of sampling needs as required

An exposure control plan is developed for one of the following purposes:

- To identify and control contaminants of concern related to worker exposure
- The ongoing monitoring needed to address either changing environmental conditions of changing contaminants (e.g., changing or addition of new chemical products in the workplace) or because the work steps of personnel performing work have changed

The plan should encompass the following at a minimum:

- Work location
- Work activity
- Similarly Exposed Group (SEG)
- Potential contaminants of concern (contaminants identified through historical and current operational research)
- Sampling type (personal or area)
- Sampling frequency (continuous, 4 or 8 hours, or every shift)
- Sampling instrument (sorbent tube or personal sampling pump)
- Data analysis and reporting
- Plan review frequency

The plan should be shared with employees once developed with a discussion on the purpose, content, and detail associated with implementing the plan to include time frame. An exposure control plan cannot be fully implemented without the engagement of workers. This is an activity where worker engagement is important for success.

14.4.2 Exposure Control Technologies

Exposure control technologies encompass the use of computer and instrumentation technologies together to provide a pictorial view that can be used to quantify and prevent exposures. In this section we will discuss some of the technologies that can and are being used successfully to aid in exposure prevention. Some of the technologies that can be used to control exposure are discussed below:

Instrumentation – Instruments that are used in monitoring contaminants continue to develop and evolve using updated and new technologies as they become available. Sensors and measurement devices are routinely utilized to evaluate suspected substances, or to conduct routine audits and surveys. Wearable instrument technology offers the ability to monitor facilities and work environments continually, as well as offering personalized insights as they are worn by individuals[3]

Sampling Medias – Sampling media are used to collect contaminants on a surface that can be analyzed to determine the presence or absence of a particular constituent of concern. There are a variety of media on the market designed to measure various types of constituents. These media are used in conjunction with other equipment such as an air sampling pump.

Computer Databases – Computer databases that track and manage exposure assessments and risks are an important technological addition to the management of risks in the protection of workers. There are several vendors that have developed computer-based programs that are widely used by a variety of companies. These databases are similar yet different in the organization of data and the presentation of input yielding information that can be helpful in risk mitigation.

TABLE 14.3
Barrier Impact Certifying Exposure Control Activities

Barrier	Barrier Impact
Lack of Employee Engagement	Employees unwilling to wear monitoring equipment
Time Constraint	Not enough available time to ensure that the date is appropriate and accurate
Budget Constraint	Inadequate funding needed to collect, and analyze the amount of data to provide credibility to the end decision
Inadequate Resources	Not enough or the appropriate type of monitoring equipment to conduct monitoring as needed
Expertise	Lack of subject matter expert knowledge to determine type of monitoring and data once collected
Different Regulations	Different regulations require different types of monitoring and analysis to certify exposures are within safe limits which requires knowledge of the interpreter when determining path forward for compliance certification
Human Factors	Change averse and not willing to engage
Lack of training	The certifier does not have the knowledge level and the credibility to certify data accuracy
Poor Safety Culture and Practices	Safety culture does not lend itself to open and honest communication that can facilitate the employee engagement needed
Work Behavior	Workers not engaged

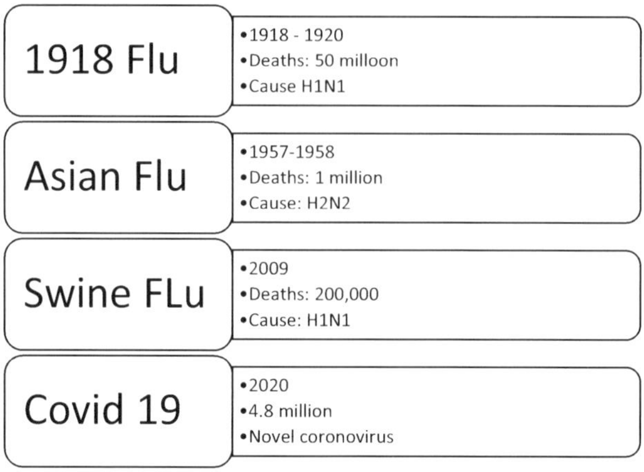

FIGURE 14.3 Pandemics of previous times[4]

14.5 PROPER ASSESSMENT AND CONTROLS DURING PANDEMICS

Pandemics are an infrequent event that creates enormous challenges for health and safety practitioners, medical communities, emergency responders, and the impacted citizens. Although not a common event, pandemics can impact a large number of people, families, and workplaces and cause significant deaths as shown in Figure 14.3.

Conducting an assessment of the work area is not easy since many of the issues encountered are considered the first of a kind. Training employees in workplace hazards is the first step in protecting workers, ensuring compliance with regulatory requirements, and ultimately reducing health risks to the workers. Risk Communication is essential in every work environment at all times; however, it may be even more essential when work is being performed during a pandemic. Workers should be made aware of the exposure risk level associated with their job duties and the work environment. A pandemic may disproportionately affect some people in certain age groups or with specific preexisting health conditions. Following the risk management steps listed in Figure 14.4 is important in identifying and managing the associated risks.

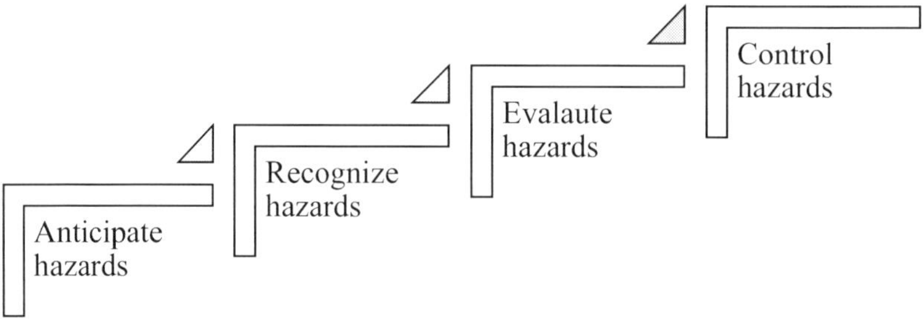

FIGURE 14.4 Risk management steps

As a part of protecting workers during pandemics, management should ensure that workers understand the following:

- The differences between seasonal epidemics and worldwide pandemic disease and the impact of outbreaks
- The risk associated with job activities and exposure to sources of infection
- What options may be available for working remotely, or utilizing an employer's flexible leave policy in the event they become ill
- Social distancing strategies to avoid close physical contact such as shaking hands and large gatherings with people
- Good hygiene sanitation procedures
- The availability and use of Personal Protective Equipment (PPE) – (how to wear, use, clean, and store it properly)
- The availability of medical services such as vaccination, post-exposure medication
- The method that will be used to communicate pandemic-related information and where to direct their questions[5]

TABLE 14.4
Barrier Impact on Proper Assessment and Control during Pandemics

Barrier	Barrier Impact
Lack of Employee Engagement	Employees not willing to follow precautionary measures
Time Constraint	Time required to determine protective measures or create vaccinations to protect people
Budget Constraint	Funds necessary to follow prescribed protective measures may be cost prohibitive for small businesses
Inadequate Resources	Obtaining the proper resources may be challenging for example a shortage or the availability of PPE
Expertise	Knowledgeable SME available to deal with pandemic fall-out
Different Regulations	New regulations or prescribed best practices may change as more information is known
Human Factors	During pandemic factors include fear and disbelief. Compliance with work practices and procedures designed to protect workers and limit the spread of a disease
Lack of training	Training may be tied to the lack of information and knowledge on the substance and or disease
Poor Safety Culture and Practices	Safety culture can facilitate acceptance of the changes in policies and practices required to ensure the safety of the workforce
Work Behavior	The behavior of workers can increase the issues surrounding working during a pandemic, for example, not wearing PPE or practicing limited contact with each other

REFERENCES

1. Koh, D., and Aw, T.-C. Surveillance in occupational health. *Occupational and Environmental Medicine (London, England)*, 60(9), 705–710, 2003, https://doi.org/10.1136/oem.60.9.705
2. Alston, F., DeKerchove, E.M., and Piispanen, W. *Industrial Hygiene: Improving Worker Health through an Operational Risk Approach*. CRC Press, 2018.
3. insights-makusafe-wearables.pdf (nsc.org)
4. History's deadliest pandemics: Plague, smallpox, flu, covid-19 - Washington Post.
5. OSHAFS- recognize, reward, or 3747.pdf

15 Resource Packet

RESOURCE PACKET

The resource packet is broken down into four sections. These sections contain resources that can be used to evaluate and improve safety and health within organizations.

Section 1 contains evaluation tools that can be effective when used appropriately to gauge the status of an organization.

Section 2 contains resources or worksheets that assist in performing hazard analysis to ensure that the hazards of the work environment and performing specific tasks are known, evaluated, mitigated, and protect workers from being harmed during performance of work.

Section 3 contains a variety of resources that can be used in assessing a safety and health program, evaluating performance, and identifying changes that may need to be instituted to improve performance. The checklist presented is a simplistic method that can be effectively used to evaluate performance.

Section 4 contains a list of resources that can be consulted in developing an effective safety and health program that will provide the level of safety for workers that is expected of employers. The list also contains resources on topics such as risk assessment, hazard analysis, and mitigation actions.

SECTION 1

Organization Culture Questionnaire

The culture questionnaire is intended to provide feedback on employee perspective with regard to the culture of their work environment.

Culture Focus Group Questions

Focus Group discussions provide an opportunity for employees to freely communicate their perspective on the culture of the organization and offer improvement measures to improve interactions between management and workers.

Employee Engagement Questionnaires

The questions listed in the survey are designed to gauge employee perspective on their engagement in the organization. Employee engagement is key to accomplishing work, continuous improvement, and facilitating a safe working environment.

DOI: 10.1201/9781032624389-19

Organization Trust Questionnaire

The questions listed in the survey are designed to determine the trust level between management and employees.

Organization Diagnostic Questionnaire Matrix – Trust

The matrix provides the connection between the survey questionnaires and the trust attributes measured providing the opportunity to determine performance by attributes.

Communication Survey Questions

Transparent communication is critical for effective organizational interactions as it establishes and facilitates trust and employee engagement. Gauging and keeping a pulse on communication is a tool that can be used to enhance trust, employee engagement, and employee confidence in the leadership team and the company. The questions listed are examples of the types of questions that can be used in a survey or focus group to gauge communication levels and areas of improvement that may need to be attended to.

Organization Culture Questionnaire									
For each question given below, circle the number that best describes your opinion of the questions listed.									
1	2	3	4	5	6	7	8		
Always	Mostly	Frequently	Usually	Sometimes	Infrequently	Seldom	Never		
#	Question	1	2	3	4	5	6	7	8
1	My organization has a clear vision								
2	The values of my organization are shared by its members								
3	I believe that my management values my opinion								
4	Communication in my organization is fluent and flows in all direction								
5	The managers in my organization recognize and celebrate the success of it's members								
6	The mission and values of my organization are posted for employees to view								
7	In my organization management celebrates the successes of employees at every level								
8	The management team is trusted and respected by employees at every level								
9	Management is responsive to suggestions from employees								
10	Conflicts are handled openly and fairly								
11	Employees are motivated to perform their jobs								
12	Employees understand their job duties and their role within the organization								

13	Downward communication is accurate								
14	The organization goals and objectives are clear to employees throughout the organization								
15	Roles and responsibilities within the organization are clear and understood								
16	My input is valued by my peers								
17	Employees have the right training and skills to perform their jobs								
18	Knowledge and information sharing is a common practice for members of my organization								
19	Disagreements are addressed promptly when they occur								
20	Morale is high across my organization								
21	Employees enjoy coming to work								
22	I feel that I am valued as a part of my team								
23	Employees speak highly of my organization								
24	Roles and responsibilities in my organization are clearly defined and understood								
25	Everyone takes responsibility for their actions								
26	My supervisor is a positive role model								

CULTURE FOCUS GROUP QUESTIONS

1. What do you like most about working for your organization?
2. What three words would you use to describe the culture of your organization?
3. What are the most common complaints workers have concerning your organization?
4. How does communication flow in your organization?
5. Are you provided the equipment and training needed to perform your job successfully?
6. How comfortable are people in your organization in communicating with management (supervisor, middle management, senior manager)?
7. Are the skills and abilities of the people in my organization valued by management and used in their current job?
8. How important is trust to the members of your organization?
9. Are employees encouraged and willing to get involved in solving issues?
10. What five words would you use to describe the management team?

EMPLOYEE ENGAGEMENT QUESTIONNAIRE

Employee Engagement Questionnaire						
For each question given below, circle the number that best describes your opinion of the questions listed.						
1	2	3	4	5		
Always	Mostly	Sometimes	Rarely	Never		
#	Question	1	2	3	4	5
1	I am willing to participate on a team to implement a program or process					
2	My manager often seeks out my opinion					
3	I often have the opportunity to participate in decisions that affects me and the organization					
4	I willingly provide feedback to my managers on issues					
5	I feel valued for the work I perform					

6	I am inspired by my management					
7	I enjoy coming to work					
8	My supervisor provides guidance when needed					
9	I am proud to tell people where I work					
10	Management provides the tools I need to do my job					
11	I understand how my role fits into the organization strategy					
12	Employees are motivated to perform well					
13	Employees in my organization take responsibility for their decisions					
14	My workgroup works together as a team					
15	I have the skills needed to do my job efficiently					
16	I am very satisfied with my job					
17	I would recommend my organization to others as a good place to work					
18	I receive the information I need to perform my job					
19	I share responsibility for the safety of myself and my coworkers					
20	I know what my supervisor expects of me while at work					
21	My coworkers and I are committed to doing a quality job					

ORGANIZATION TRUST QUESTIONNAIRE

Organization Trust Questionnaire						
For each question given below, circle the number that best describes your opinion of the questions listed.						
1	2	3	4	5		
Always	Mostly	Sometimes	Rarely	Never		
#	Question	1	2	3	4	5
1	Senior management in my company communicates information completely and frequently					
2	Managers in my company support a work-life balance for all workers					
3	The managers in my organization are not afraid to admit when they are wrong					
4	Clear and concise communication permeates the organization					
5	The managers in my organization keep the promises they make					
6	Disagreements and issues are addressed in a timely manner					
7	The leaders in my organization are decisive in their decision making					
8	Management provides clear, concise, and honest feedback to subordinates and colleagues					
9	The managers in my organization demonstrate good leadership and management skills while conducting business and making decisions					
10	Information is communicated in totality, without holding back critical elements that are important for employees to know					

11	My workplace is safe and clean
12	People in my organization treat each other with respect
13	Organization members are willing to share information
14	My company's benefit program is adequate and comparable to similar companies
15	The managers in my organization lead with confidence
16	The manager in my organization listens attentively to workers
17	Organizational members are receptive to new ideas
18	Communication flows in all directions in my organization to ensure that workers are kept informed
19	People in my organization follow policies and procedures
20	The policies and practices in my organization ensure that workers are compensated fairly
21	My organization is able to retain the skilled workers needed to complete work

Overall Score: ________

Attribute Scores:

Openness and Honesty ______ Competence ______ Identification ______

Reliability _______ Relationship _________ Concern for Employees _______

ORGANIZATION DIAGNOSTIC QUESTIONNAIRE MATRIX – TRUST

1. Senior management in my organization communicates information completely frequently (openness and honesty)
2. Managers in my company support a work-life balance for all workers (concern for employees)
3. The managers in my organization are not afraid to admit when they are wrong (openness and honesty, relationship)
4. Clear and concise communication permeates the organization (openness and honesty)
5. The managers in my organization keep the promises they make (reliability)
6. Disagreements and issues are addressed in a timely manner (relationship)
7. The leaders in my organization are decisive in their decision making (competence)
8. Management provides clear, concise, and honest feedback to subordinates and colleagues (openness and honesty)
9. The managers in my organization demonstrate good leadership and management skills while conducting business and making decisions (competent)
10. Information is communicated in totality without holding back critical elements that are important for employees to know (openness and honesty)
11. My workplace is safe and clean (concern for employees)
12. People in my organization treat each other with respect (relationship)
13. Organization members are willing to share information (openness and honesty)
14. My company benefit program is adequate and comparable to similar companies (concern for employees)

ORGANIZATION DIAGNOSTIC QUESTIONNAIRE MATRIX – TRUST - CONTINUED

15. The managers in my organization lead with confidence (competence)
16. The managers in the organization listen attentively to workers (relationship)
17. Organization members are receptive to new ideas (openness and honesty)
18. Communication flows in all directions in my organization to ensure that workers are kept informed (concern for employees)
19. People in my organization always follow policies and procedures (identification)
20. The policies and practices in my organization ensure that workers are compensated fairly (concern for employees)
21. My organization is able to retain the skilled workers needed to complete work

QUESTIONS THAT CAN BE INCLUDED IN A COMMUNICATION SURVEY OR ROUND TABLE

Rate your comfort in using communication channels to share feedback or concerns.

1. Very comfortable
2. Comfortable
3. Neutral
4. Uncomfortable
5. Very uncomfortable

How often do you feel well-informed about company updates?

1. Always
2. Often
3. Occasionally
4. Rarely
5. Never

On a scale of 1 to 5, how would you rate the overall communication experience at our company?

1. 1 (Poor)
2. 2 (Below average)
3. 3 (Average)
4. 4 (Above average)
5. 5 (Excellent)

How satisfied are you with the overall level of internal communication?

1. Very satisfied
2. Satisfied
3. Neutral
4. Unsatisfied

I feel free to express my ideas to upper management.

1. Always
2. Often
3. Occasionally
4. Rarely
5. Never

Do you feel that management listens to you when you speak?

1. Yes
2. No
3. Frequently
4. Infrequent

Do you understand the Goals of the company?

1. No
2. Yes

The team is kept well informed by management on what's going on in the company.

1. Always
2. Often
3. Occasionally
4. Rarely
5. Never

Do you believe that company leaders always tell the truth about company performance?

1. Always
2. Often
3. Occasionally
4. Rarely
5. Never

Discussion Questions

- How do you get information about the company and impending changes?
- How can the company improve communication with employees?
- What can your management team do to improve communication?
- How would you describe the level of communication across the company?

SECTION 2

Job Hazard Analysis (JHA) Worksheet – JHA is an effective procedure that when performed adequately is effective in integrating safety and health procedures into task performance through the identification, elimination, and/or control of the associated hazards. The worksheet provides an example of completing the process and the types of analysis that should be included in the process.

Exposure Assessment Worksheet – An effective way to identify hazards in the workplace and hazards associated with performing various tasks.

Personal Protective Equipment (PPE) Hazard Assessment Certification and Selection Worksheet – An evaluation of the hazards and work conditions should be performed to determine the type of PPE that should be prescribed. The PPE Worksheet provides an avenue to perform and document that evaluation and decision made regarding PPE type.

Air Sampling Worksheet – Often there is a need to sample the work environment to determine the presence or absence of hazardous air constituents. When the need arises, use an air sampling worksheet to ensure collection of samples and document environmental conditions while collecting samples as well as instrument data.

Exposure Assessment Report – Once the work area has been assessed or evaluated it is important to communicate the results gleaned to the workforce. An example of the content of an informative report is provided to guide documentation and communication of results.

Air Sampling Report – The air sampling report is designed to consolidate information such as the sampling result to include data, exposure summary information, pre- and post-instrument sampling calibration records, laboratory results and field notes, and recommendations that may be made as a result of the information obtained.

EXPOSURE ASSESSMENT DATA COLLECTION WORKSHEET

Exposure Evaluation/Assessment Data Collection Worksheet

Date:_______________________________ANALYTICALLABORATORY

Evaluation Conducted by: _____________________________________

Location: ____

Task/Activity: ____

____

Physical Hazards of concern:

Chemical Hazards of concern:

Qualitative Evaluation: (list in this section data that include procedures, policies etc. reviewed)

Quantitative Evaluation: (list data that may have been previously collected)

Personnel Interviews:

Name: ____________________ Title: ____________ Date: ___________

Name: ____________________ Title: ____________ Date: ___________

AIR SAMPLING DATA SHEET

Sample Date: ___________ Sample Number __________ Sample Type: _______

Worker Name:	
Worker Job Category:	
Work Location:	

WORK AND ENVIRONMENTAL CONDITIONS

Wind Speed: __________ Wind Direction: _________ Humidity: ________ Temperature: _____

Work Activity: ____

____

____

Process Description: ____

____

Hazards: ____

____

____

PPE: ____

____

EQUIPMENT CALIBRATION DATA

Sampling Pump Model and Mfg:	Flow Rate:	Units:	Sample Start Time:
Sampling Pump Serial Number:	Start:		Sample End Time:
	End:		Duration in minutes:
	Average:		Volume in liters:

Calibrator Mfg. & Model	Calibrator Serial number	Annual Calibration Date

SAMPLING RESULTS (ANALYSIS)

Substance (s)	**Concentration & Units**	**Exposure Limit**

ANALYTICAL LABORATORY

Date:	Analyzed by:	Calculated by:	Date:	Checked by:	Date:

PERSONAL PROTECTIVE EQUIPMENT (PPE) HAZARD ASSESSMENT CERTIFICATION AND SELECTION

Work Location: ____

____

Work Task: ____

____

____

Assessment completed by: ____

Assessmentdate:____

Noise Hazard: (noise hazards greater than 85 dBA)

Hazard	Description of Hazard	Mitigation Actions	Required PPE
Machinery			
Electrical (arc flash)			
Rotating Equipment			

Eye Hazard:

Hazard	Description of Hazard	Mitigation Actions	Required PPE
Chemical			
Dust			
Light/radiation			
Electrical (arc flash)			
impact			

Hand Hazard:

Hazard	Description of Hazard	Mitigation Actions	Required PPE
Chemical			
Cut/abrasion			
Hot/cold exposure			
Heat-burns			
Puncture			

Foot Hazard:

Hazard	Description of Hazard	Mitigation Actions	Required PPE
Chemical			
Compression			
Impact			
Puncture			

Fall Hazard:

Hazard	Description of Hazard	Mitigation Actions	Required PPE
Confined space			
Heights greater than or equal to 4 feet			
Working at heights			
Operating industrial vehicles			
Slippery surfaces			
Unprotected edges			

Respiratory Hazard:

Hazard	Description of Hazard	Mitigation Actions	Required PPE
Vapors			
Dust			
Respirable fibers			
Gas			
Fume			
Mists			

Certified by:

Name: ____

Signature: ____

Date: ____

SAMPLE JOB HAZARD ANALYSIS (JHA)

Task Analyzed: ____

Work Location: ____

Supervisor: _________________________________ Date: ______________

Sequence of Tasks	Potential Hazards	Mitigation Actions	Required PPE
	1. 2. 3.	1. 2. 3.	

JHA PARTICIPANTS

Name	Work Group	Signature	Date

Completed by: ________________ Signature: ____________ Date: _________

SECTION 3

These resources/checklists contained in this section can be used as examples and are not inclusive of all actions or activities that should be taken to ensure a healthy safety and health program. Included are:

- Assessment Checklists
  - Facility Basic Safety Conditions
  - Slips Trips and Fall Hazards
  - General Office Safety
  - Hazard Communication

    - Safety & Health Training
    - Respiratory Protection
    - Confined Space
    - Medical Surveillance
    - Ergonomics
    - Occupational Noise
- Assessment Plan Example
- Example of Criterion and Lines of Inquiries (LOIs)
- Improvement Plan (Example)

FACILITY BASIC SAFETY CONDITIONS

Organization:	**Building:**	**Room/Location:**	**Completed by:** **Date:**

Check N/A if the item does not apply.

	Yes	No	NA	Comments
Walkways are free from obstructions				
Access to the exit is unobstructed and doesn't require travel through a high-hazard area				
Stairways are in good condition with handrails and non-slip treads in good condition				
Exit signs are visible and illuminated				
Stairway steps are uniform in size				
Exit doors closed and not braced open				
Fire extinguishers are easily accessible, safety seals in place, checked monthly, and operational				
Fire extinguishers mounted so that the travel distance from the employee work area and an extinguisher is 75 feet or less				
Adequate lighting is in the work areas and walkways				
Fire extinguisher inspection tag has been initialed within the past 30 days of the last inspection (or within the month)				

OSHA posters are displayed in places that can be viewed easily				
Electrical panels are secured and unobstructed (minimum clearance in front of the panel 36 inches and 30 inches horizontally)				
Eye wash stations are in locations where chemicals are used and are tested annually				
PPE is provided, accessible maintained in good condition, and stored appropriately				
Housekeeping is maintained and all equipment and supplies are in proper storage				
SDSs are available for office and housekeeping chemicals				

SLIPS TRIPS AND FALL HAZARDS

Organization:	**Building:**	**Room/Location:**	**Completed by:** **Date:**

Check N/A if the item does not apply.

	Yes	**No**	**NA**	**Comments**
Floors free of liquids (e.g., oil and water)				
Floor surfaces are level and intact				
Floors are clean, dry, free of trip hazards, and in good condition				
Aisles and walkways are unobstructed				
General housekeeping conditions are maintained				
Cords and cables are not draped along the floor				
Permanently used cords are covered when crossing walkways/work areas				
Mats and carpeting are affixed to the flow (no curling edges)				
All physical hazards such as cracks, holes, etc. are identified until they can be repaired				

GENERAL OFFICE SAFETY

Organization:	Building:	Room/Location:	Completed by: Date:

Check N/A if the item does not apply.

	Yes	No	NA	Comments
Furniture is free from damage and corners and edges are smooth or protected				
Office items are stored in a stable manner, not subject to falling				
Electrical cords are not located where they become tripping hazards.				
Electrical cords are protected against accidental damage, with no daisy chaining extension cords or use of multi-outlet strips				
Portable heaters are not located near paper products, upholstered furniture, or in confining spaces				
Portable heaters are on level surfaces and plugged directly into wall outlets.				
Filing cabinets secured eliminating the risk of falling or tipping over				
Filing cabinets and desk drawers are closed when not in use				
Conference room occupancy evaluated marked				
Workstations have been evaluated for ergonomic risks to eliminate or reduce risk factors				

HAZARD COMMUNICATION

Organization:	Building:	Room/Location:	Completed by: Date:

Check N/A if the item does not apply.

	Yes	No	NA	Comments
A hazard communication policy is in place				
Each hazardous chemical used in the work area has a Safety Data Sheet (SDS)				
SDSs are located in an area where they are immediately available to workers				
Hazard communication labels present and legible on containers, tanks, etc., where hazardous materials are stored				
Target organ effects are listed on each hazard warning label				
There is an audit/inspection system to periodically check for labels that may have fallen off or become illegible				
OSHA's Form 300A is displayed conspicuously in occupied workspaces				
Hazardous Communication training is provided to employees				
The training includes information on processes in the work areas where hazardous chemicals are present				
Training includes information on the OSHA Hazard Communication Standard and the requirements of the standard				
Training includes measures employees can take to protect themselves from chemical hazards				
Chemical spill response kits are stocked and available				
Compressed gas cylinders are labeled, stored upright, and are secured				
Incompatible materials are segregated (e.g., acids and bases)				
Flammable gas cylinders are stored away from oxidizing gas cylinders				
Flammable liquids are stored in flammable cabinets				

SAFETY & HEALTH TRAINING

Organization:	Building:	Room/Location:	Completed by: Date:

Check N/A if the item does not apply.

	Yes	No	NA	Comments
Fire extinguisher training is conducted upon initial assignment and updated annually				
Hazardous Communication training is provided to employees				
Workers receive training in a language and at a literacy level that they can understand and retain				
Workers receive supplemental training when a change in the work environment and task structure could introduce new or increased hazards				
Workers are trained to report injuries and changes in workplace conditions promptly				
Managers and supervisors understand their responsibilities under the OSHA Act				
Managers and supervisors understand procedures for responding to workers' reports of injury, illness, or concern				
Managers and supervisors understand techniques for identifying and controlling hazards in the workplace				

RESPIRATORY PROTECTION

Organization:	Building:	Room/Location:	Completed by: Date:

Check N/A if the item does not apply.

	Yes	No	NA	Comments
Hazard assessment completed that identifies tasks/ activities that require respiratory protection to be worn to include the type				
Respirators in inventory are approved by NIOSH				
Workers are medically evaluated prior to wearing respirators				
Workers are fit-tested to wear a respiratory protection device				
A program is in place to clean and disinfect respiratory protection devices				
A program is in place to maintain respiratory protection devices to ensure protection factor is retained				
Employees are trained on proper inspection and wearing or respiratory protection devices that they wear				
The respiratory program administrator is engaged in program administration				
The program is evaluated annually by the program administrator				

CONFINED SPACE

Organization:	**Building:**	**Room/Location:**	**Completed by:** **Date:**

Check N/A if the item does not apply.

	Yes	No	NA	Comments
Is the space large enough for a person to enter to perform work				
Is the space designed for continuous occupancy				
Does the space have restricted means to enter and exit the space				
Is a special tool required to access the space				
Have workers received the appropriate confined space training				
Has an emergency plan been developed and in place				
Has a supervisor been appointed responsibility for the space (tasks)				
Is mechanical ventilation required to enter the space				
Has supervisor received training and understands the hazards associated with confined spaces				
Are confined spaces posted				
Is there a written confined space program				
Are confined space entry permits used to enter spaces				
Are the spaces evaluated and tested for hazards				

MEDICAL SURVEILLANCE

Organization:	**Building:**	**Room/Location:**	**Completed by:** **Date:**

Check N/A if the item does not apply.

	Yes	No	NA	Comments
A comprehensive medical surveillance program to identify health hazards, develop health and safety programs, and increase the health and safety of employees is in place				
The effectiveness of a medical surveillance program is continually reviewed and upgraded based on new technologies and changes in regulatory requirements				
Employees are trained in medical surveillance requirements for their workplace				
Employees undergo regular medical examinations for as long as they work with toxic, radioactive, or other hazardous materials in the workplace				
Detailed records of workers' medical examinations from pre-employment to the latest medical exam is kept				
There is an enrollment and disenrollment process based on exposure potential and data				
The Medical Director oversees the Medical Surveillance program				
The Medical Director works in conjunction with EH&S to identify the hazardous health exposures and place the employee into the appropriate surveillance program				
Employees have access to medical records upon request				

ERGONOMICS

Organization:	**Building:** **Person Observed:**	**Room/Location:**	**Completed by:** **Date:**

Check N/A if the item does not apply.

	Yes	No	NA	Comments
Work activities have been evaluated for ergonomic risk factors				
Have there been any worker complaints regarding ergonomic-related issues				
Employees perform repetition tasks				
High-impact tools used routinely				
Employees use awkwardly designed tools, which cause the worker to operate the tool outside of a neutral position for long time period				
High vibration producing tools used routinely				
Routine tasks require repeated heavy-lifting				
There is adequate lighting in the workplace				
The workstation is configured in such a way that prevents undue stress on the body				
Employees receive training in ergonomic risk factors and mitigation actions				
Employees take frequent breaks when working high stressed activities				
A written ergonomic program is in place and implemented				

OCCUPATIONAL NOISE

Organization:	**Building:**	**Room/Location:**	**Completed by:** **Date:**

Check N/A if the item does not apply.

	Yes	No	NA	Comments
Workplace areas have been assessed for noise				
Areas exceeding 85dDA are posted				
Hearing protection devices provided and available				
Audiometric testing available and provided				
Control methods using the hierarchy of controls are incorporated into the program				
Hazard identification and exposure monitoring are conducted				
Employees are trained to select, use, and maintain devices				
Is medical surveillance available for employees exposed to noise above 85 dBA				
Are there any employees with a Standard Threshold Shift				
Equipment needed to measure noise available, maintained, and calibrated				

ASSESSMENT PLAN EXAMPLE

Scope:

The scope outlines the purpose of the assessment and what areas that are expected to be included in the assessment.

The Assessment Team

Documents the team members, their area of responsibility, experience, and expertise. It also includes the assigned area that the team member will be responsible for assessing when there is a multi-topic assessment being performed.

The Assessment Method

This section of the Plan describes the methods that will be used to demonstrate and document the outcome and how it was achieved.

Data Collection

Documents the methods that are used during the assessment. Methods can include a review of previous assessments, regulatory actions, completed checklists, management assessments, interviews, etc. The timeframe for data collection should be included.

Results Reporting

The purpose and the goals and the outcome expected from the assessment.

EXAMPLE OF CRITERION AND LINES OF INQUIRIES (LOIS)

1. Lines of Inquiry for implementation of an effective Hearing Conservation Program
   Criteria: The hearing conservation program is effective in preventing employees from suffering hearing damage from overexposure to noise above 85 dBA
   a. LOI 1 – A document hearing conservation program is in place
   b. LOI 2 – If information indicates that any employee's exposure may equal or exceed an 8-hour time-weighted average of 85 decibels, the employer has developed and implemented a noise monitoring program
   c. LOI 3 – The monitoring program identifies employees for inclusion in the hearing conservation program and to enable the proper selection of hearing protectors
   d. LOI 4 – Instruments used to measure employee noise exposure are calibrated to ensure accurate measurement
   e. LOI 5 – Employee exposed at or above an 8-hour time-weighted average of 85 decibels of the results of the monitoring are promptly notified
   f. LOI 6 – An audiometric testing program has been provided and audiometric testing is available to all employees whose exposures equal or exceed an 8-hour time-weighted average of 85 decibels
   g. LOI 7 – Within 6 months of an employee's first exposure at or above the action level, the employer has established a valid baseline audiogram that will be used to compare subsequent audiograms
   h. LOI 8 – At least annually after obtaining the baseline audiogram, a new audiogram is obtained for each employee exposed at or above an 8-hour time-weighted average of 85 decibels
   i. LOI 9 – Employees are fitted and trained in the use of hearing protectors and provided with hearing protection devices
   j. LOI 9 – Hearing protection devices are available to employees at no cost

k. LOI 10 – Employees wear hearing protection devices when working in areas that are at or above 85 decibels
l. LOI 11 – An accurate record of all employee exposure measurements is retained
m. LOI 12 – all employee audiometric records are maintained

2. Lines of Inquiry for assessment of a respiratory protection program for compliance

   Criteria: A written respiratory protection program has been developed and implemented a written with worksite-specific procedures and elements for required respirator use.

   a. LOI 1 – Procedures for selecting respirators for use in the workplace are available
   b. LOI 2 – Evaluations of employees that are required to use respirators are medically evaluated
   c. LOI 3 – Procedures and schedules for cleaning, disinfecting, storing, inspecting, repairing, discarding, and maintaining respirators are in place
   d. LOI 4 – Respirator training is provided at no cost to the employee
   e. LOI 5 – Respiratory hazard(s) in the workplace has been evaluated and the results are available to employees
   f. LOI 6 – Medical evaluations are provided at no cost to the employee
   g. LOI 7 – The employer has identified and secured a physician or other licensed health care professional (PLHCP) to perform medical evaluations
   h. LOI 8 – The PLHCP determines the employee's ability to use a respirator
   i. LOI 9 – Employees are fit tested with the same make, model, style, and size of respirator that will be used before wearing a respirator
   j. LOI 10 – Respirators are stored to protect them from damage, contamination, dust, sunlight, extreme temperatures, excessive moisture, and damaging chemicals, and packed or stored to prevent deformation of the facepiece and exhalation valve
   k. LOI 11 – Respirators are inspected before each use and during cleaning
   l. LOI 12 – Effective training is provided to employees who are required to use respirators
   m. LOI 12 – Written information regarding medical evaluations, fit testing, and the respirator program are retained

IMPROVEMENT PLAN – EXAMPLE

Issue (s)	Corrective Action (s)	Start Date	Completion Date	Responsible Resource
Not all workers requiring hazard recognition training received training	Identify workers requiring training Provide training to workers that had not received training	00/00/0000	00/00/000	John Doe
The respiratory protection procedure does not address the frequency of respiratory protection retraining	Update the respiratory protection procedure to address the time frame for retraining of workers	00/00/0000	00/00/000	John Doe
Respirators are improperly stored in a circular trash bin in an area on the shop floor that is not climate controlled	Insect respirators in storage to ensure that they are not damaged Identify proper storage location relocate respirators to new storage location	00/00/0000	00/00/000	Jane Doe
Employees are not always fitted and trained in the use of hearing protectors	Evaluate tasks requiring employees to wear respirators Identify employees requiring fit tested and training Provide training and fit test employees	00/00/0000	00/00/000	Jane Doe
Confined spaces are not identified, evaluated, or posted as required	Identify all confined space Conducted an evaluation of spaces Post spaces requiring identification by posting as a confined space	00/00/0000	00/00/000	Jane Smith

SECTION 4

OTHER RESOURCES

1. Guidelines for Evaluating an EPA Partnership Program (Interim), guidelines-eval-https://www.epa.gov/sites/default/files/2015-09/documents/guidelines-eval-epa-partnership-programs.pdfepa-partnership-programs.pdf
2. Recommended Practice for Safety And Health Programs, Safety Management - https://www.osha.gov/safety-management/program-evaluationProgram Evaluation and 3. Improvement | Occupational Safety and Health Administration (osha.gov)
3. Safety and Health Program Self-Evaluation Tool for General Industry, https://www.osha.gov/sites/default/files/SHP_Self-Evaluation_Tool.pdf
4. https://www.ccohs.ca/oshanswers/hsprograms/checklist/list_mft.html
5. What is program evaluation, A beginner's guide, last revised 3/10/2020, https://www.evaluation.gov/assets/resources/Resource_What%20is%20Program%20Evaluation%20-%20A%20Beginners%20Guide.pdf
6. https://www.directives.doe.gov/directives-documents/400-series/0414.1-EGuide-1C/@@images/file
7. Air Sampling in the Workplace, US Nuclear Regulatory Commission, September 1993, ml13051a671.pdf (nrc.gov)
8. Air Sample Report, https://www.dir.ca.gov/DOSHPol/p&pc-91.pdf
9. Guidelines for Human Exposure Assessment, Environmental Protection Agency https://www.epa.gov/sites/default/files/2020-01/documents/guidelines_for_human_exposure_assessment_final2019.pdf
10. George Gruetzmacher, 82 AIHA Exposure assessment strategy for assessing industrial hygiene exposures, *Annals of Work Exposures and Health*, Volume 68, Issue Supplement_1, June 2024, Page 1, https://doi.org/10.1093/annweh/wxae035.266

Index

A

Active errors, 45
Acts of vulnerability, 176
Adjustable workstations, 165, 166
Administrative controls, 94, 96, 115, 117, 121, 122, 125, 127–130
The Alienated, 20
American Industrial Hygiene Association (AIHA), 79, 80, 103
Appropriate controls, 89, 132
Artificial intelligence (AI), 161
Assessment attributes, 150
Assessment documentation, 155
Assessment in brief, 156
Assessment out brief, 157
Assessment plan, 34, 151–153, 156, 195, 218, 227
Assessment strategy, 149, 151, 153, 194
Assessment team, 35, 92, 151–154, 156, 227
Avalanche rescue, 120, 129, 130
Aversion therapy, 49, 50

B

Barriers to safety & health, 187
Behavioral reinforcement, 51
Behavior based, 47
Behavior modification, 43, 45, 47–51, 53, 54, 59
Body language, 63, 70, 71, 73–75, 77, 80
Bomb squad, 120, 125, 126
Border guard, 120, 126, 127
Budget constraints, 188

C

Classification of Gig workers, 104
Coast Guard, 120, 127
Common hazards and descriptions, 98, 99
Communication with employees, 74, 211
Communication with senior leaders, 75
Communication with supervisors, 75
Communication with the occupational medicine provider, 75
Complacency, 45, 46, 53
Concern for employees, 58, 59, 74, 208, 209
Continuous improvement, 57, 64, 135, 142–144, 146, 194, 201
Continuous improvement cycle, 143
Control methods, 117, 126–130, 227
COVID-19, 198
Creativity, 16, 55, 57, 58, 61, 62
Cultural attributes, 14, 55, 56
Culture improvement plan, 69
Customer satisfaction, 58, 61, 63, 133
Custom workstation considerations, 167
Custom workstations, 166, 167

D

Data analysis, 35, 94, 142, 190, 197
Data collection, 11, 34, 67, 80, 94, 141, 142, 152, 170, 212, 228
Data communication, 73
Data Interpretation, 94
Data reporting, 72
Department of Labor, 30, 100, 184
Developing leadership skills, 18
Different regulations, 111, 189, 192, 195, 197, 198
Disruptive behaviors, 46
Documentation of assessments, 155

E

ECG tests, 164
Effective strategy, 147
Emergency medical services (EMS), 114–117, 119, 121, 130
Emergency Medical Services Safety & Health, 121
Emergency responders, 114, 117, 123, 124, 198
Emergency responders burnout, 123
Emergency response, 127, 162, 172
Emergency service worker, 114, 115, 117, 119–121, 123–125, 127, 129
Employee engagement, 8, 60, 61, 64, 68, 71, 133, 135, 138, 140, 142, 187, 188, 192, 195, 197, 199, 201, 202, 205
Employee involvement, 90, 163, 187
Employee value, 59
Employment market evolution, 102
Engineering controls, 94, 117, 121, 122, 125, 128–130, 192
Environmental health, and safety (EHS), 162
Epidemiologist, 10, 11
Ergonomists, 10, 11
Error likely, 43
Error precursors, 52, 53

Error types, 45
Evaluating organizational culture, 64
Example JHA, 97
Exoskeletons, 163, 164
Expertise, 118, 148, 149, 188–192, 195, 197, 199
Exposure assessment, 91, 190, 193, 194, 196, 197, 212
Exposure control, 190, 191, 194–197
Exposure control plan, 195, 196
Exposure control technologies, 196
Exposure monitoring, 77, 78, 191, 192, 194, 227
Exposure Surveillance, 190–192
Extinction, 49, 50

F

Fall protection, 129
Federal Aviation Administration, 171
Firefighters, 114–117
Flooding, 49, 50
Followers, 8, 19, 20, 23, 24
Followership type, 20

G

Gig employees safety & health, 104
Gig workers, 103–107
Gig workers categories, 106
Good communicator, 71–73
Good listener, 73

H

Hazard analysis, 70, 87–89, 91–97, 99, 100, 125, 133, 135, 180, 191, 195, 201, 211, 217
Hazard communication, 14, 89, 217, 220, 221
Hazard identification, 7, 76, 88–90, 149, 227
Hazardous material, 119, 120, 125, 129, 149, 185, 187, 221, 225
Health monitoring, 76, 193
Health surveillance, 190
Hidden workers, 101
High performance, 55, 57, 58, 64
Human behavior, 44, 55, 56
Human characteristics, 45, 53
Human factors, 11, 188, 189, 192, 195, 197, 199
Human performance, 43–45, 47, 52, 55, 138, 178
Human Performance Improvement Program (HPIP), 52

I

Implementing an assessment plan, 152
Improved performance, 60
Improvement implementation plan (IIP), 157, 158
Inaccurate view of risk, 45, 46
Inadequate resources, 189, 192, 195, 197, 199
Inadequate resources, 189, 192, 195, 197, 199
Independent contract employees, 111
Independent contractors, 104, 105, 109, 111, 112
Industrial hygienist, 10, 11, 76, 78–80, 82, 191
Innovation, 4, 21, 57, 58, 61–63, 161
Inspection checklist, 155, 156

J

JHA data collection, 94
JHA preventive action, 96
Job hazard analysis, 87–89, 92–94, 97, 99, 100, 125, 135, 211, 217
Job hazard analysis process redesign, 92
Job hazard analysis program, 93
Job Hazard Analysis software, 97
Job safety analysis, 94, 99, 125
Job Safety and Environmental Analysis (JSEA), 92, 93
Job selection, 89

K

Key performance indicator, 34, 136, 137
Knowledge sharing, 62

L

Lack of training, 189, 192, 195, 197, 199
Lagging indicators, 99, 132, 133, 136
Language barrier, 175, 176, 179
Latent errors, 45
Laws and regulations, 28–34, 36, 39, 44, 102, 147–149, 175, 183, 184
Leadership development, 18
Leadership ethics, 23–25
Leadership focused roles, 4
Leadership role, 4, 67
Leadership selection, 16
Leading indicators, 132, 133, 136
Lean thinking, 57, 60
Lessons learned, 92, 93, 144–146
Lumbar motion monitors, 164

M

Management engagement, 58, 155
Management role, 3, 4, 6
Measuring performance, 125, 131, 133, 134
Metric, 131, 132, 134, 137–141
Migrant worker, 176, 180, 182

N

National Fire Protection Association (NFPA), 116, 117
Negative reinforcement, 49, 50
Non-traditional employee, 101
Non-traditional work, 101, 102

O

Occupational health nurse, 10, 11
Occupational health physician, 76
Occupational Medicine Physicians, 10, 11, 76
Occupational safety & health, 29, 31, 116, 125, 130, 184–186, 192
Occupational Safety & Health Administration (OSHA), 31, 39, 88, 99, 116, 147, 184
Organization culture, 67, 68, 142, 201, 203
Organization structure, 5–7, 60, 148
Overworked employees, 181

P

Pandemics, 198, 199
Park rangers, 120, 127, 128
People aspects, 5–7
Performance measures, 132–135, 141, 142
Performance metrics, 133, 134, 141
Personality traits, 46, 53
Personal protective equipment (PPE), 14, 88, 89, 94, 96, 106, 121–123, 149, 161, 185, 187, 192, 199, 212, 214
Person-Organization Fit Assessment, 123
Physiological status monitor, 164
Policy setting, 6, 7
Poor Safety Culture, 190, 192, 195, 197, 199
Positive reinforcement, 49, 50
The Pragmatic, 20
Pregnant women, 175, 177
Pregnant Woman Fairness Act, 177, 186
Prioritizing work tasks, 95
Probability classification, 95
Procedure development, 6, 7, 76
Process review and analysis, 91
Product and service cost, 63, 64
Product and service quality, 64
Program assessment strategies, 153
Promoting from within, 16, 17
Psychologists, 10, 11
Punishment, 48–50

R

Reduce recall of information, 46
Regulatory manager, 147
Remote work, 103, 107, 108
Resource packet, 64, 157, 201
Risk assessment, 11, 125, 165, 193–195
Risk management, 28, 140, 198

S

Safety & health program, 5, 22, 31, 39, 70, 75, 76, 116, 137, 138, 141, 147, 149–151, 156, 187, 190, 201, 217
Safety & health program areas, 149
Safety & health technicians, 11
Safety culture, 7, 116, 163, 167, 179, 188, 190, 192, 197, 199
Safety Data Sheet (SDS), 89, 98, 221
Safety leader, 8–10, 12–14, 26
Safety leadership, 8–10, 13
Safety performance, 8, 32, 51, 132, 140, 141, 146
Safety professional, 7, 10, 11, 78
Safe Work Method Statement (SWMS), 92, 93
Search and Rescue responders, 128
Selecting leaders, 16, 21, 22, 26
Sensors, 162, 164, 170, 197
Severity classification, 95
Sharing of knowledge, 61–63
The Sheep, 20
Short term memory loss, 46
Situational awareness, 46
SMART, 136, 137
Specialized emergency services, 119, 124
Stages of a Job Hazard Analysis, 89
The Star Follower, 20
Step trackers, 164
Stress, 12, 30, 45, 46, 53, 99, 116, 117, 120, 123, 124, 129, 164, 167, 180, 189, 193
Strong communicator, 71
Succession strategy, 17
Systematic desensitization, 49

T

Task breakdown, 89
Teamwork, 61, 62, 168
Telework, 107, 109
Telework Employees Safety & Health, 107
Time constraint, 187, 188, 192, 195, 197, 199
Tone of voice, 74, 77, 80
Top talent, 57, 58
Toxicologists, 11
Training courses, 14
Training delivery methods, 13
Transformational leader, 8, 9
Types of workstations, 167

U

United States Fire Administration, 117
Use of drones, 169–171

V

Visualization technology, 172
Vulnerable workers, 175, 176

W

Wearable device functions, 183
Wearable technology, 161–164
Why Assess, 147
Work control, 92
Work culture, 8, 23, 45, 47, 58, 65, 68, 149, 181, 187
Worker behavior, 90, 94, 104, 190
Worker burnout, 180, 182
Workers with disabilities, 177
Workforce health monitoring, 76
Workplace equity, 183
Workplace laws, 30, 31, 50, 175, 183–186
Workplace violence, 105, 117, 121, 122, 182
Workstation safety, 169
Written communication, 78, 81, 82
Written communication, 78, 81, 82

Y

The Yes Person, 20

Made in the USA
Las Vegas, NV
19 September 2026

11497fb1-d6e1-4abe-98f1-55424f6676fcR01